Rudolf Neumann

Two-Scale Thermomechanical Simulation of Hot Stamping

Schriftenreihe
Kontinuumsmechanik im Maschinenbau
Band 11

Karlsruher Institut für Technologie (KIT)
Institut für Technische Mechanik
Bereich Kontinuumsmechanik

Hrsg. Prof. Dr.-Ing. habil. Thomas Böhlke

Eine Übersicht aller bisher in dieser Schriftenreihe erschienenen Bände finden Sie am Ende des Buchs.

Two-Scale Thermomechanical Simulation of Hot Stamping

by
Rudolf Neumann

KIT Scientific Publishing

Dissertation, Karlsruher Institut für Technologie
KIT-Fakultät für Maschinenbau

Tag der mündlichen Prüfung: 30. März 2017
Referenten: Prof. Dr.-Ing. Thomas Böhlke
Prof. Dr.-Ing. Dr.-Ing. E.h.A. Erman Tekkaya

Impressum

KIT Scientific Publishing

Karlsruher Institut für Technologie (KIT)
KIT Scientific Publishing
Straße am Forum 2
D-76131 Karlsruhe

KIT Scientific Publishing is a registered trademark
of Karlsruhe Institute of Technology.
Reprint using the book cover is not allowed.

www.ksp.kit.edu

Print on Demand 2017 – Gedruckt auf FSC-zertifiziertem Papier

ISSN 2192-693X
ISBN 978-3-7315-0714-7
DOI 10.5445/KSP/1000073431

Two-Scale Thermomechanical Simulation of Hot Stamping

Zur Erlangung des akademischen Grades

Doktor der Ingenieurwissenschaften

der Fakultät für Maschinenbau

Karlsruher Institut für Technologie (KIT)

genehmigte

Dissertation

von

Dipl.-Ing. Rudolf Neumann

Tag der mündlichen Prüfung:	30.03.2017
Hauptreferent:	Prof. Dr.-Ing. Thomas Böhlke
Korreferent:	Prof. Dr.-Ing. Dr.-Ing. E.h. A. Erman Tekkaya

Zusammenfassung

Die thermomechanischen Eigenschaften von Stahl werden signifikant von der thermomechanischen Behandlung während eines Fertigungsprozesses beeinflusst. Unterschiedliche physikalische Zustände des Stahls, genannt Phasen, können dadurch erzeugt werden. Die Komposition dieser Phasen in der Mikrostruktur legen das effektive thermomechanische Verhalten des Stahls fest. Presshärten ist ein thermomechanischer Tiefziehprozess, der dieses Verhalten des Stahls nutzt, um Bauteile mit einem guten Festigkeit-Gewicht-Verhältnis zur Erhöhung der Insassensicherheit und Treibstoffreduktion von Fahrzeugen zu erzeugen.

Das Hauptziel dieser Arbeit ist die Formulierung, die numerisch effiziente Implementierung, und die Anwendung eines zweiskaligen Modells für die Simulation des Presshärteprozesses. Dazu wird ein nichtlineares Homogenisierungsschema vom Hashin-Shtrikman-Typ derart erweitert, dass temperaturgetriebene Phasenumwandlungen und die phasenumwandlungsbegleitenden Effekte berücksichtigt werden. Dabei wird im diffusionsgesteuerten Fall die Phasenumwandlung durch ein erweitertes JMAK- oder Kirkaldy-Venugopalan-Model und im diffusionslosen Fall durch das Koistinen-Marburger- oder einem "S"-Shape-Model beschrieben. Die Umwandlungsdehnung wird als isotrop angenommen. Die transformationsinduzierte Plastizität wird durch einen verallgemeinerten Ansatz nach Leblond umgesetzt. Das thermomechanische Phasenverhalten wird durch eine Helmholtz-Energie beschrieben, die aus den Annahmen einer konstanten Wärmekapazität, einer von der plastischen Verformungsgeschichte unabhängignen elastischen Eigenschaft, und einer linearen Spannungsrelation abgeleitet wird. Zusätzlich wird die thermodynamische Konsistenz des Gesamtmodells bestehend aus freier Energie und Evolutionsgleichungen für die inneren Variablen nachgewiesen.

Nach einer Parameteridentifikation anhand von experimentellen Ergebnissen wird das zweiskalige Model dazu verwendet, Dilatationsversuche und Presshärteprozesse von Stahl zu simulieren. Dabei wird das thermomikromechanische Modell anhand von experimentellen und numerischen Ergebnissen eines Referenzmodells validiert.

Summary

The thermomechanical properties of different types of steel are significantly influenced by the thermomechanical treatment during a production process. Different physical states of the steel, referred to as phases, can be obtained. The composition of these phases on the microstructure determines the effective thermomechanical behavior of the steel. Hot stamping is a thermomechanical deep drawing process which takes advantage of this steel behavior to produce parts with a good strength-to-weight ratio to increase the occupant security and to reduce the fuel consumption of vehicles.

The main objective of this work is to formulate a nonlinear two-scale model, to realize a numerically efficient implementation of this model, and to apply the model for the simulation of a hot stamping process. Thereby, a nonlinear Hashin-Shtrikman-type homogenization scheme is extended to take the temperature-driven phase transformation effect and the phase transformation accompanying effects into account. The phase transformation is described by an extended JMAK or Kirkaldy-Venugopalan model in the diffusion-driven case and by the Koistinen-Marburger or a suggested "S" shape model in the diffusionless case. The transformation strain is assumed to be isotropic and the transformation-induced plasticity is modeled by a generalization of Lebond's model. The thermomechanical behavior of the phases is described by a Helmholtz energy derived by the three assumptions of a constant specific heat capacity, elastic properties unaffected by plastic effects, and linear thermomechanical stress response. In addition, the thermodynamic consistency of the complete model consisting of free energy and evolution equations for the internal variables is shown.

After a parameter identification on the basis of experimental results, the derived two-scale model is applied to simulate both the dilatation process and the hot stamping process of sheet metal. Thereby, the thermo-micromechanical model is validated via experimental results and numerical results of a reference model usually used in literature.

Acknowledgements

Many people had a direct or indirect influence on the present thesis. At this point, I want to thank these people for the explicit or implicit support which made the success of my work at the Chair for Continuum Mechanics possible. Hereby, I would like to express my gratitude to my supervisor Prof. Dr.-Ing. Thomas Böhlke who already got me enthusiastic about mechanics during my studies at the University of Kassel and who offered me the possibility to work in his research group at the KIT. I enjoyed the fruitful professional and private discussions and I am grateful for his support, guidance, and commitment.

Furthermore, I would sincerely like to thank Prof. Dr.-Ing. Dr.-Ing. E.h. A. Erman Tekkaya for the co-review of my thesis. With his valuable remarks and the possibility of accomplishment of hot stamping experiments during a research stay at the IUL in Dortmund, he contributed significantly to the success of my work.

In this context, I would like to thank the members of the Graduate School 1483 for endless discussions, providing experimental results, and experimental support, in particular, from the IAM-WK in Karlsruhe and the Frauenhofer IWM in Freiburg. My special thanks go to the German Research Foundation (DFG) for funding my project within the Research Training Group GRK 1483.

Further, I highly acknowledge the support of the Karlsruhe House of Young Scientists (KHYS) for funding my research stay abroad at the Georgia Institute of Technology in Atlanta. I highly appreciate that Prof. David L. McDowell invited me to the GeorgiaTech's George W.

Woodruff School of Mechanical Engineering to experience three fruitful and memorable months in his research group.

My thanks also go to my colleagues at the Institute of Engineering Mechanics (ITM) for five years of sharing countless unforgettable moments in science and in private matters. In particular, chewing the fat with Viktor in the office, with Barthel in the Kani-Restaurant, with Vedran in the Kanu, Mauricio in Peru, Eric during the breaks of our intensive Squash matches, and with my other colleagues during the distractive coffee breaks are going to stay with me forever. I am glad that I found not only colleagues but also friends.

The great time that I had at the ITM, would not have been possible without the careful planning and support of Helga, Ute, and Tom, who also receive my special thanks. You helped me a lot to focus on my scientific work by taking many organizational tasks off me.

Last but not least, I would like to thank my family for their unconditional support of my studies and my friends outside of the scientific field for the distraction and for a wonderful time in Karlsruhe.

Contents

Chapter 1

Introduction

1.1 The Hot Stamping Process

The hot stamping process was invented by the Swedish company Plannja in 1977 (Karbasian, 2010). Since the first use of hot stamped parts in automotive production in 1984 by the Swedish company Saab, the amount of hot stamped parts manufactured per year has increased up to over 400 million parts which are manufactured in 2013 (see Figure 1.1). In automotive production, hot stamped parts are used as rear and front bumpers, central reinforcements, or A and B pillars. Due to the characteristics of the manufacturing, hot stamped parts combine the advantages of high strength and low weight components. The good strength-to-weight ratio contributes to both the reduction of fuel consumption and an increase of occupant safety (see, e.g., Karbasian and Tekkaya, 2010; Merklein and Lechler, 2006). It is for this reason that crash-relevant components in automotive industry are often manufactured by hot stamping.

The hot stamping process is a hot working, deep-drawing process and is usually performed in two different ways: direct and indirect hot stamping. The direct hot stamping process consists of three main steps (see on the left-hand side of Figure 1.2): the semi-finished sheet metal is

1. Austenitized at about 1200K,
2. Transferred to the deep-drawing press, and

3. Formed and quenched in the closed deep-drawing press.

The austenitization of the sheet metal leads to a good formability and a high fracture strain (see the fracture-strain-tensile-strength-diagram of the steel grade 22MnB5 on the right-hand side of Figure 1.2). The hot forming step followed by quenching in the closed press delivers parts with a high tensile strength of up to 1500MPa which are manufactured in one go. If the sheet metal is initially cold formed to a net shape of 90%

Figure 1.1: Development of the amount of hot stamped parts per year in automotive (Billur and Altan, 2015)

before hot stamping is performed, the process is referred to as indirect hot stamping. Indirect hot stamping is used for complicated geometries which cannot be deep drawn in one go (see, e.g., Hippchen et al., 2016). The advantages and disadvantages of the hot stamping process are the

- Manufacturing of parts with a good strength-to-weight ratio (tensile strength up to 1500MPa) in one step,
- Low springback effect after unloading of the final part (see, e.g., Bao et al., 2013; Yanagimoto and Oyamada, 2005),
- Small forming forces due to forming of austenitized sheet metals,

- High energy demand due to thermal treatment of the sheet metal, and
- Difficult further treatment of hot stamped parts.

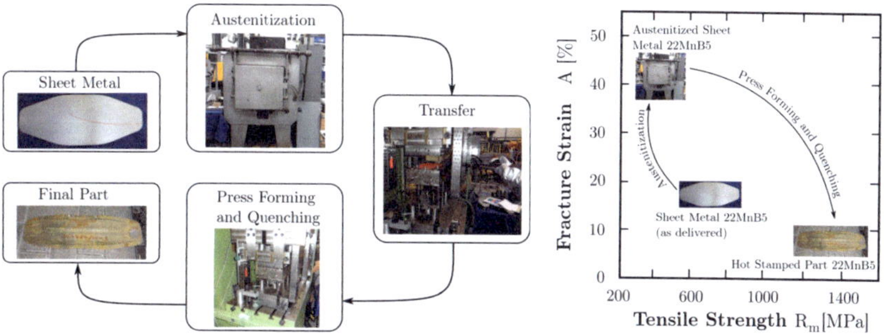

Figure 1.2: Process chain of the hot stamping process (left) and the hot stamping process in the fracture-strain-tensile-strength-diagram (right)

To overcome the difficulties of the hot stamping process, i.e. the high energy consumption and difficult further treatment, new heating methods beyond the conventional furnace heating and different coating technologies have been proposed. For the austenitization of sheet metals, induction and conduction heating have been investigated by, e.g., Bariani et al. (2008); Kollecka et al. (2009) and Mori et al. (2005; 2015); Liang et al. (2015), respectively. Conduction heating is highly efficient but results in a inhomogeneous temperature distribution for complex part geometries. The efficiency of the induction heating is dependent on both the distance between the coil and the sheet metal and on the temperature of the sheet metal. Beyond the Curie temperature, the efficiency of the induction heating decreases with temperature until the austenitization temperature is reached (Karbasian, 2010).

In hot stamping, the coating protects the sheet metal against scaling, decarburization and corrosion (Fan and De Cooman, 2012), as well as improving the tribological properties of the sheet during forming.

Fan and De Cooman (2012) reviewed the different coating methods used in hot stamping based on aluminized, Galvanized, Galvannealed, Zn-Ni alloy, and hybrid coatings. Borsetto et al. (2009) investigated the Al-Si coating in hot stamping conditions and found that, in contrast to the holding time, the cooling velocity has no influence on the coating structure nor on the chemical composition. The shorter the holding time the more homogeneous the coating structure appears at the expense of a higher roughness. Due to a low forming capacity, the Al-Si coating is not applicable for indirect hot stamping. Furthermore, the Al-Si coating does not provide cathodic protection, which is desirable for hot stamped parts (see, e.g. Karbasian and Tekkaya, 2010).
For these reasons, the Zinc based coatings are used to improve the tribological parameters and the cathodic protection. Furthermore, due to the high forming capacity, Zn-Ni coated sheet metals are able to be used in a prior cold forming step before the hot stamping takes place (see, e.g., Cho et al., 2014; Kondratiuk et al., 2011; Seok et al., 2015). Kondratiuk and Kuhn (2011) compared the tribological friction and wear behavior of Al-Si and Zn-Ni coatings. They found that the Zn-Ni coating provides a lower friction coefficient but, due to ZnO formation, an enhanced wear of the workpiece compared to the Al-Si coating. However, due to adhesive wear characteristics of Al-Si coatings, the mass transfer and persistent material build-up on the die is significantly lower if Zn-Ni coatings are used.
A third nanotechnology-based coating system is the x-tec coating. For further information, see Goedicke et al. (2008) and Karbasian (2010).

The austenitized sheet metal has to be transferred to the press as quickly as possible to reduce the loss of temperature. The fast transfer ensures both the good formability of the austenitized sheet metal at a high temperature, as well as no prior phase transformation which would lead to crack formation during the forming step (see, e.g., Karbasian, 2010). Abdulhay et al. (2011) discussed the cooling of the heated part during transfer due to convection and radiation. Lenze and Sikora (2006) investigated the cooling process in air and gave guide values for the cooling rate of the sheet metal dependent on the sheet metal thickness.

To ensure that the forming of the austenitized sheet metal takes place before any phase transformation occurs, the forming has to be performed with a high stamp velocity and a short forming time. In the closed press, a high heat transfer is desired to get cooling rates above the critical cooling rate (CCR), at which a pure martensitic microstructure is obtained. As Abdulhay et al. (2012), Bosetti et al. (2010), Caron et al. (2014), Kim et al. (2015), and Merklein and Lechler (2006) have shown, the magnitude of the heat transfer coefficient of the contact between the tool and the sheet metal is dependent on the pressure. They found that the higher the pressure, the greater the heat transfer coefficient. For this reason, in order to avoid prior phase transformation and cracks in the flange of the sheet metal, the blank holder is distanced from the sheet metal. The distance between blank holder and sheet metal is dependent on the geometry. Due to the complexity of the relationship, the distance is often determined by the trial-and-error method in industrial applications (see, e.g., Karbasian, 2010). For blank holder distances above a critical value, wrinkles occur in the flange, whereas for blank holder distances under a critical value, cracks occur in the part.

To produce high strength parts, quenching has to take place at a cooling rate above the CCR. In his work, Naderi (2007) investigated different material types to find the best steel to produce a pure martensitic microstructure in a common hot stamping process. He found, that only the steel grades 22MnB5, 27MnCrB5, and 37MnB4 produce a pure martensitic microstructure in a process with a water- or nitrogen-cooled punch. Steel grades without boron develop ferrite and/or bainite during hot stamping. Adding Boron as an alloying element slows down the decomposition of the austenitic phase into ferrite and bainite by segregation at the grain boundary (see, e.g., Gárlipp et al., 2001; Jun et al., 2006).

Furthermore, Naderi (2007) found that 37MnB4 has the greatest and 22MnB5 the lowest strength with a tensile strength up to 2040MPa and 1493MPa, respectively. However, as measured by the formability index UTS A25, the formability of 22MnB5 is twice as good as of 37MnB4, if a water-cooled punch is used, and three times higher, if a nitrogen-cooled punch is used.

Depending on the local cooling rate of an austenitized sheet metal, different microscopic states are obtained. The ability of steel to occur in different micromechanical states, either face-centered-cubic or body-centered-cubic lattice or the kind of carbon dissolution, is referred to as polymorphic behavior. Regions of equal chemical composition and thermomechanical properties on the microstructure are referred to as phases. Usually, four different states which can arise from austenite during quenching can be distinguished: ferrite, pearlite, bainite, and martensite. While ferrite and pearlite arise due to diffusion of carbon on the micro scale, martensite occurs in a diffusionless reorientation and distortion of the austenitic lattice. In the latter case, the cooling takes place beyond the CCR at which no diffusion of carbon can occur. The bainitic phase arises in moderate cooling where both diffusion of carbon and the reorientation and distortion of the austenitic lattice are of equal importance. Due to complex geometries and locally different thermomechanical conditions, the cooling rates are locally different. Thus, sheet metals with graded material behavior or tailored properties can be produced (see, e.g., Abdollahpoor et al., 2016). Graded material properties are desired to improve the properties of the final part in view of the safety performance by enhancing local energy absorption of the component during a crash (see, e.g., Wang et al., 2014). To realize graded material behavior of a sheet metal, four different methods are mostly used (see, e.g., Hochholdinger, 2012; Karbasian and Tekkaya, 2010; Merklein et al., 2016):

- Local annealing of hot stamped parts (see, e.g., Labudde and Bleck, 2009; Shi et al., 2012; Wang et al., 2014),
- Local heating and austenitization of the sheet metal before hot stamping (see, e.g., Liang et al., 2014; Liu and Lei, 2015; Mori, 2012; Mori et al., 2013; Wang et al., 2014),
- Locally influenced heat transfer by locally heated tools (see, e.g., George et al., 2012) or locally different tool materials (see, e.g., Casas et al., 2008), and
- Hot stamping of tailor welded blanks (see, e.g., Kinsey et al., 2000; Lechler et al., 2010; Tang et al., 2007; 2013).

While, in the first three methods, the tailored properties are obtained by adjusting the microscopic structure of 22MnB5, the hot stamping of tailor-welded blanks needs a complementary material and a prior joining step. Appropriate materials have been investigated in several experimental trials by Lamprecht and Deinzer (2008), Stopp et al. (2007), and Tang et al. (2013).

By whatever means hot stamped parts are produced, be it by direct or indirect hot stamping, the process is the interaction of three different physical phenomena: mechanics, heat conduction, and phase transformation. The interaction of these three parts is depicted in Figure 1.3. The mechanical properties of a part are determined by the microstructure which evolves during the hot stamping process. The different thermomechanical properties of the phases influence the heat transport in the bulk. Furthermore, during the evolution of the microstructure, energy is consumed or emitted which has an effect as a sink or a source for the energy balance. In addition, the mechanical properties of a single phase are temperature-dependent, which affects the forming behavior of the bulk. During the process, irreversible inelastic forming takes place which goes hand-in-hand with a release of energy referred to as mechanical dissipation. A strong coupling between the mechanical and phase transformation effect occurs due to the transformation-induced plasticity.

The transformation-induced plasticity (TRIP) effect plays an important role in hot stamping processes. In the case of a phase transformation under external load, the non-classical plasticity effect takes place even if the external load implicates stresses far beneath the yield stress of the bulk. Two different effects cause TRIP strains: the Magee and the Greenwood-Johnson (GJ) mechanism. In Cherkaoui et al. (1998) and Taleb and Sidoroff (2003) these two effects are defined as follows:

- "The Greenwood-Johnson mechanism corresponds to the micromechanical plastic strain arising in the parent phase from the expansion of the product phase",
- "The Magee mechanism corresponds to the formation of the selected martensitic variants resulting from the applied stress".

Due to this, the Magee effect represents the combination of an orientation effect of the austenitic single crystal and the corresponding strain arising during the transition from a face-centered-cubic (FCC) lattice of austenite to a body-centered-cubic (BCC) (see, e.g., Bhadeshia, 1985; Guo, 2004) or body-centered-tetragonal (BCT) (see, e.g., Turteltaub and Suiker, 2005) lattice of martensite (see, e.g., Magee, 1969). Bain and Dunkirk (1924), Kurdjumov and Sachs (1930), Nishiyama (1934), Wassermann (1935), Greninger and Troiano (1949), and Pitsch (1967) introduced reorientation relations for the parent austenitic lattice during the transformation to the martensitic lattice which led to 24 possible martensitic variants in the Kurdjumov-Sachs model. Nolze (2004) investigated the orientation relation between the FCC and BCC lattice in the context of dual steels. He found that, in some cases, the experimental orientation relation is not always well-described by one of the orientation relation models but is an intermediate state between them. During phase transformation, the characteristics of the external and internal load determines which variants are energetically advantageous and which are more likely to arise (see, e.g., Cherkaoui et al., 2000; Levitas and Preston, 2002; Ostwald et al., 2010). In polycrystalline materials, this selection of growing variants leads to an oriented martensitic phase with a macroscopic permanent distortion due to the anisotropic Bain strain of the advantageous variants. This distortion is reversible which is a characteristic of the shape memory effect (SMA) (see, e.g., Ganghoffer and Simonsson, 1998).

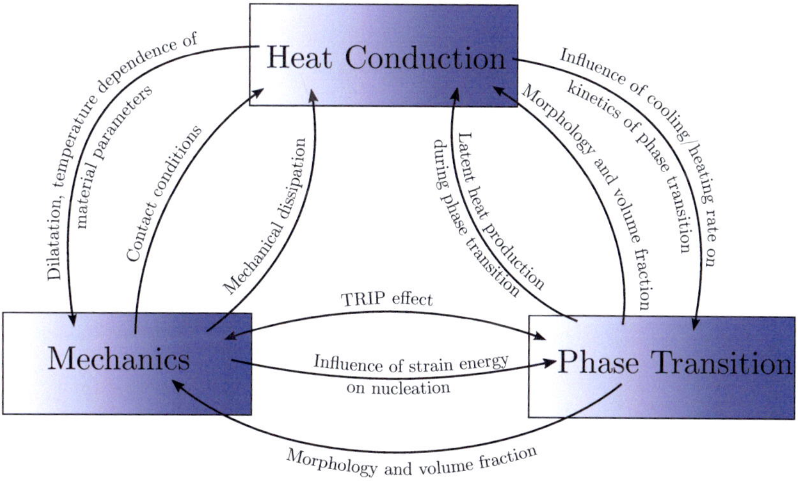

Figure 1.3: Interaction diagram of the mechanical problem, the heat conduction, and the phase transformation

In contrast to the Magee effect, the GJ mechanism describes the plastification of the austenitic phase during the growth of martensite. The difference in the thermal strains of austenite and the growing phase, the change in volume of the austenitic phase during transformation, and the inhomogeneous distribution of nuclei on the micro scale yield to deviatoric stresses in the austenitic phase beyond the yield strength. The microscopic plastic strains add up in a measurable macroscopic irreversible distortion (see, e.g., Leblond et al., 1989; Taleb and Sidoroff, 2003).

During a phase transformation under external load, the GJ and Magee effect arise simultaneously. However, depending on the thermomechanical boundary conditions, either the GJ or the Magee effect play a greater role for the TRIP distortion. Sierra and Nemes (2008) found that the GJ effect has no significant contribution to the TRIP effect in the case of the forming of a multi-phase steel with retained austenite at room temperature. In such conditions, the austenite has a relatively high yield strength and hardly plasticizes during phase transformation.

However, during the hot stamping process, the phase decomposition of austenite takes place at elevated temperatures. The austenite has a low yield strength and almost the entire austenitic phase plasticizes as shown by Leblond et al. (1989). In such a case, the GJ effect is the main contributor to the TRIP distortion. Therefore, the Magee effect is neglected in hot stamping-related works (see, e.g., Ackerström et al., 2007; Bok et al., 2014; Durrenberger et al., 2009; Hochholdinger et al., 2011; Kim et al., 2012; Wu et al., 2015).

1.2 Simulation of the Hot Stamping Process - State of the Art

1.2.1 Challenges on the Modeling of the Hot Stamping Process

Due to the increasing importance of the hot stamping process, which is reflected by the continuous increase of the production figures per year since 1987 (see Figure 1.1), the demand on simulation tools capturing the complex thermomechanical characteristics of the process has become necessary for optimization and development purposes. Neugebauer et al. (2012) have shown the current challenges in the hot stamping processes which are the

i. Determination of suitable process parameters to ensure a high safety process,
ii. Minimization of resources by optimized component geometry and energy consumption reduction,
iii. Realization of energy efficient tools, machinery, and equipment,
iv. Optimization of the process with regard to material or energy loses, and
v. Optimization of the cycle time.

With accurate numerical simulations, the safety of the process, the geometry and properties of hot stamped components, and necessary cooling times for an optimized downtime can be predicted in an early stage of the development of a hot stamping process. To ensure a high accuracy, a simulation tool should be able to

1. Capture the complex thermomechanical boundary conditions of the process including a contact formulation,
2. Describe the pressure and gap dependence of the heat transfer through the surfaces in contact,
3. Describe the diffusion-driven and the diffusionless phase transformation of ferrite, pearlite, bainite, and martensite in the parent phase,
4. Take the temperature and phase-dependent elasticity, plasticity, and heat conduction into account,
5. Consider viscous effects in elasticity and plasticity,
6. Capture the TRIP effect occurring when phase transformation takes place under external and internal load, and
7. Consider the finite rotation and distortion of the material points in the forming step of the austenitized sheet metal.

1.2.2 Models Suggested in Literature

In the context of hot stamping, Ackerström et al. (2007), Olsson (2009), Hochholdinger et al. (2011), and Bok et al. (2014) proposed thermomechanical models and implemented them into either the explicit finite element (FE) simulation tool LS-Dyna (2016) or the implicit FE simulation tool of ABAQUS (2016). The works use a phenomenological approach by modeling the strain rate and temperature-dependent yield stress of austenite with a semi-empirical ansatz of Hoff (1954)- or Johnson and Cook (1983)-type as well as a physically-based ansatz of Nemat-Nasser (1999). The other phases are assumed to be rate independent, and modified models by Voce (1948)-, Swift (1952)-, or Hockett and Sherby (1975)-type are used. In these approaches, the overall plastic behavior is obtained by the simple mixture rule applied on the yield stresses of the different phases. In such an approach, only one internal variable describing the isotropic plasticity exists. Since almost the entire plasticity in hot stamping takes place in the austenitic phase, due to forming of the austenitized sheet or the GJ effect, this approach appears to be reasonable with the assumption that the entire dislocation density is taken by the growing phases during phase transformation. Furthermore, the overall thermal properties, i.e. the conductivity and the heat capacity, are also determined by a simple mixture law.

In the cited works, however, the elastic properties of the bulk are assumed to be independent of the microstructural composition. For example, Bok et al. (2014) used Young's modulus for boron steel as a function of temperature as proposed by Turetta et al. (2006). Indeed, the experimental results for the austenitic and the martensitic phase, which are the hot stamping determining phases, show a deviation in the Youngs's modulus of martensite from austenite of about 26% at a temperature of 650K. For the same phases at the same temperature, the thermal expansion coefficients differ by up to about 50% (see, e.g., Miokovic, 2005; Schwenk, 2012).

Beyond the models which are directly aimed at simulating the hot stamping process, Wolff et al. (2008b) and De Oliveira et al. (2010) proposed thermomechanical multiphase (TMMP) models based on a thermodynamic framework. In a small deformation context, Wolff et al. (2008b) introduced an overall Helmholtz free energy and De Oliveira et al. (2010) a Helmholtz free energy for each consisting phase. Under suitable conditions, these procedures yield to thermodynamically consistent material models. While they modeled the plasticity in the same way as the hot stamping directed works, i.e. in a phenomenological manner with one internal variable for all phases, the elastic and thermal parameters have been taken into account by a simple mixture law which is an upper first-order bound for the elasticity as well as for the conductivity.

1.2.3 Modeling of the TRIP Effect

Since, in hot stamping, the phase transformation takes place under external load, TRIP occurs and contributes to the residual stress development. The application and opportunities, the material behavior, and the constitutive modeling of SMAs, which are mainly determined by the Magee effect, are outlined well in Jani et al. (2014), Lobo et al. (2015), and Cisse et al. (2016), respectively. For further discussion see, e.g., Achenbach (1989), Bartel and Hackl (2009), Brinson (2004), Buchelnikov and Bosko (2003), Christ and Reese (2009), Govindjee and Hall (2000), Helm and Haupt (2003), Husson et al. (2011), Kastner et al. (2011), Ostwald et al. (2012), Panico and Brinson (2007), Wang et al. (2008), and Yu et al. (2014a).

The GJ effect observed by Greenwood and Johnson (1965) has been heuristically modeled by Denis et al. (1985), Desalos et al. (1982), Mitter (1987), and Sjöström (1985). Leblond et al. (1989) and Leblond (1989) underpinned the heuristically derived equation for the evolution of the macroscopic TRIP strain by micromechanical arguments. While in Leblond's approach the growing phase is rigid, Taleb and Sidoroff (2003) reevaluated his approach considering the case that both phases are elastic. This leads to an extended model which agrees better with experimental results and which avoids singularities at the beginning of the transformation.

Since this type of model, which is well reviewed by Fischer et al. (1996), has been derived for a two phase composite, in hot stamping, where several phases can arise, Ackerström et al. (2007) and Olsson (2009) extended the model by introducing an overall hard phase variable taking the development of all phases into account. However, Wolff et al. (2008b), De Oliveira et al. (2010), and Bok et al. (2014) proposed an overall TRIP strain consisting of the sum of each phases' TRIP strain given by Leblond's equation. Based on that approach, Neumann and Böhlke (2013) and Böhlke et al. (2014) proposed a model in the context of thermo-micromechanical (TMM) modeling. Based on the work of Cherkaoui et al. (1998), another approach was proposed by Kubler et al. (2011) who introduced and evaluated a mean instantaneous transformation strain by energetically motivated arguments. As a special case, the resulting model includes Leblond's equation and, moreover, is able to take the Magee effect into account.

In the context of finite deformations, for the description of the overall TRIP effect, Oberste-Brandenburg and Bruhns (2004), Papatriantafillou et al. (2006), Mahnken et al. (2012) suggested a TRIP Green-Lagrange strain, dilatation rate, or velocity gradient, respectively, based on Leblond's model.

1.2.4 Modeling of the Phase Transformation Effects

During the cooling process, two different mechanisms trigger phase transformations in steel: the diffusion of carbon with an reorientation of the austenitic lattice (see, e.g., Bhadeshia, 1985) and the diffusionless reorientation of the austenitic lattice accompanied by Bain strain. In literature, the diffusion-driven phase transformation is usually modeled by the

- Physically based phenomenological approach of Johnson-Mehl-Avrami-Kolmogorov (JMAK) (see, e.g., Kolmogorov, 1937; Avrami, 1939; Johnson and Mehl, 1939),
- Phenomenological approach of Kirkaldy-Venugopalan (KV) (see, e.g., Kirkaldy and Venugopalan, 1983), or
- Approach based on internal variables combined with thermodynamical arguments (see, e.g., De Oliveira et al., 2010; Moumni et al., 2011; Mahnken et al., 2015)

For thermomechanical simulations, the JMAK equation, reviewed in Fanfoni and Tomellini (1998), has been used by, e.g., Lee and Lee (2008), Caseiro et al. (2011), De Oliveira et al. (2010), and Bok et al. (2014). The physically based approach of JMAK is premised on the description of the phase transformation determining mechanisms of nucleation, growing, and impingement.

In classic nucleation theory (see, e.g., Kolmogorov, 1937; Johnson and Mehl, 1939; Avrami, 1939; Jones and Bhadeshia, 1997; Kooi, 2006), a distinction is made between subcritical embryos and supercritical germs. The subcritical embryos consume energy to be able to grow, whereas the supercritical germs release energy during the growth process. The transition of subcritical embryos into a supercritical germ is referred to as nucleation. Three fundamental types of nucleation are presented in Liu et al. (2007): the case of saturation by pre-existing supercritical germs, continuous, and avrami nucleation. For the description of diffusion-driven phase transformation, the continuous nucleation is used in several applications (Todinov, 1998; Miokovic, 2005; Caseiro et al., 2011; Bok et al., 2011). Assuming a constant sum of subcritical embryos and supercritical germs, the avrami nucleation determines the nucleation rate by the rate of transition from subcritical embryos into supercritical germs. Dependent on this transition rate from subcritical embryos into supercritical germs of the phase transformation, the saturation, the maximum number of supercritical germs, is reached after a certain nucleation time.

Two growth modes of the nucleated germs can be distinguished: interface controlled growth and volume diffusion controlled growth. In the case of interface controlled growth, diffusion of atoms in the vicinity of the transformation front dominate the phase transformation. For example, the diffusion-driven phase transformation of austenite to ferrite is determined by the interface controlled growth mechanism (Bhadeshia, 1985). In contrast, volume diffusion controlled growth is influenced by long range diffusional changes of the composition. In early research, Zener (1946) and Hillert (1957) assumed the growth rate of pearlite to be volume diffusion controlled. By this assumption, the theories that have been developed underestimate the growth rate of pearlite leading to later research taking the interface controlled growth into account (Hillert, 1975; Bos and Sietsma, 2007; Pandit and Bhadeshia, 2011). This mixed formulation of the growth rate results in better agreement with measurements.

The importance of the adequate description of the impingement of the developing phases is shown in, e.g., Pradell et al. (1998). In Pradell et al. (1998), the influence of the impingement type on the phase transformation is shown. Investigating the soft and the hard impingement, Pradell et al. (1998) found that, in the case of crystallization of amorphous alloys, the assumption of soft impingement during phase transformation leads to a good agreement with experimental results. The case of existing grain boundaries of the supercritical germs with no interaction between the growing germs is referred to as hard impingement. In contrast, the case of overlapping diffusion fields of the individual germs during phase transformation is called soft impingement (Offerman et al., 2002).

With the assumptions that the

- Initial state is homogeneous,
- Phase transformation takes place under isothermal conditions,
- Subcritical embryos are randomly distributed,
- Growth is isotropic and spatially independent, and
- Hard impingement due to blocking effect is negligible

Kolmogorov (1937), Avrami (1939), and Johnson and Mehl (1939) independently derived the well-known evolution equation for diffusion-driven transforming phases. Anisotropic growth and hard impingement of the growing germs have been discussed by, e.g., Starink and Zahra (1998), Starink (2001), Kooi (2006), and Liu et al. (2007). By the introduction of an impingement determining exponent in the growth equation, they derived an evolution equation for the diffusion-driven transforming phase in which the JMAK equation is present as a limit case. Todinov (2000) discussed some limitations of the JMAK equation and found that the "equation predicts correctly the real transformed fraction only if the number of the growing nuclei in the controlled volume is large". Usually, in modeling thermomechanical processes, the physical JMAK parameters are not chosen, but rather the heuristically motivated parameters (see, e.g., Visintin, 1987; Hömberg, 1996; Miokovic, 2005; Lee et al., 2009; Caseiro et al., 2011; Schwenk, 2012; Böhlke et al., 2014; Bok et al., 2015). The aim of the phenomenological approach is to capture accurately the temperature-time-transformation (TTT) and the continuous-cooling-transformation (CCT) diagrams to be able to perform accurate thermomechanical simulations.

Since the JMAK equation and the KV model are only valid for phase transformations under isothermal conditions, for the description of the phase transformation in the non-isothermal case, a step model compliant to the Scheil's rule (Scheil, 1935) is applied. Originally, the Scheil's rule was derived to estimate the incubation time to the start of the ferritic phase transformation. The idea is to decompose the thermomechanical non-isothermal process in isothermal steps of a certain time length. The sum of the time fractions defined by the time available for the transformation by the time needed for the transformation at the temperature level is one. In the limit Scheil's rule in integral form is got. Lusk and Jou (1997), Todinov (1998), Rios (2005), and others have discussed the validity of Scheil's additivity rule if the JMAK equation is used. They found that the validity is given only for a constant Avrami exponent. Assuming a non-constant Avrami exponent, Lusk and Jou (1997) showed that the resulting CCT diagram deviates slightly from the resulting CCT diagram if an associated evolution law valid for the additivity rule is used.

Réti and Felde (1999) and Rios (2005) extended or generalized the Scheil's additivity rule. Based on the generalized rule, they were able to derive new evolution equations and, in some cases, to get better agreement with experimental results.

Another extension of the JMAK equation has been performed by Bok et al. (2015). In their work, based on the work of Kamamoto et al. (1985) and Boyadjiev et al. (1996), Bok et al. (2015) proposed a modification of the JMAK parameter toward a function of the temperature and current cooling rate. In such an approach the transformation start and finish temperatures are no longer constant. For the steel 22MnB5, the CCT curve has been captured accurately and the hardness and volume fractions of the quenched specimens have been validated for a wide range of cooling rates.

In the context of hot stamping, Ackerström et al. (2007), Olsson (2009), and Hochholdinger et al. (2011) used the KV model to describe the evolution of the diffusion-driven phase transformation. Based on the Zener-Hillert model, (Zener, 1946; Hillert, 1957), Kirkaldy and Venugopalan (1983) derived an equation to model and predict the incubation time of the diffusion-driven transformation. They calibrated their model to various TTT curves in the US Steel Atlas (see, e.g., United States Steel

Company Research Laboratory, 1963). Since the original KV model underestimates the hardenability of a metal, Li et al. (1998) modified the model by changing the reaction rate and the effect of alloying elements, and calibrated their model to CCT diagrams. To capture the influence of boron on the transformation behavior, Ackerström and Oldenburg (2006) extended the reaction rate of Li's model. Compared to the JMAK model, the KV model offers the following benefits: first, the evolution equation is only determined by the chemical composition, and second, the KV model takes the average grain size of the austenitic phase into account. In experimental investigations of the steel 22MnB5, Sun and Zhang (2013) found that the grain size of the austenitic phase differs depending on the heating method, heating rate, and heating temperature. In an ideal hot stamping process, the homogenized austenite is formed before phase transformation takes place. The plastic deformation of the austenitic microstructure and the temperature change, cause a change in the grain refinement. Based on the works of Suehiro et al. (1987) and Yoshie et al. (1996), Bok et al. (2014) introduced a dislocation based effective austenite grain size concept into the KV model.

The third class of models describing the diffusion-driven phase transformation is the thermodynamically consistent derived evolution law. In a thermodynamical framework, the volume fractions of the growing phases appear as internal variables. Introducing a potential for the dissipation or its dual, a generalized standard material approach provides the evolution law for the determination of the volume fractions (see, e.g., De Oliveira et al., 2010; Moumni et al., 2011; Mahnken et al., 2015). De Oliveira et al. (2010) introduced a dissipation potential in accordance with the JMAK model.

For high cooling rates, in particular quenching, the change in microstructure is determined by distortion of the former austenite FCC lattice into a BCC/BCT martensite lattice. The high temperature change does not allow a diffusion of iron and alloys. Reaching the martensite formation start temperature (MS), the phase transformation is governed by the undercooling below this temperature. At a certain temperature level below the MS, the transformation of the austenitic to a martensitic lattice takes place with speed of sound until the equilibrium of the chemical and mechanical energies is reached. At this point, further transformation is only obtained by further undercooling.

For isothermal processes, Koistinen and Marburger (1959) found by accurate experimental results that the logarithm of the volume fraction of austenite is proportional to the undercooling below the martensite formation start temperature (see, e.g., Christian, 2002). This leads to the well-known Koistinen-Marburger (KM) model valid for the isothermal transformation. Inoue and Wang (1985) extended the KM model by an exponential parameter for the influence of the undercooling. Furthermore, Hougardy and Yamazaki (1986) proposed a MS-dependent relation of the KM parameters. The MS is dependent on the

- Carbon and alloying elements concentration in the austenitic phase (see, e.g., Wang et al., 2000; Pitsch and Sauthoff, 1992; Andrews, 1965),
- Stress state on the microstructure (see, e.g., Todinov et al., 1996; Alexander et al., 1995),
- Austenite grain size (see, e.g., Capdvila et al., 2003; Yang and Bhadeshia, 2009), and
- Cooling rate of the austenitic specimen (see, e.g., Nikravesh et al., 2012).

In their work on the prediction of the MS via artificial neural networks (ANN), Vermeulen and van der Zwaag (1996) collected several models usually used in literature. The models differ up to 80K for the boron steel 22MnB5. In general, the most alloying elements reduce the MS. In a hot stamping process, diffusion of carbon can occur, leading to a lower carbon concentration in the remaining austenitic phase. This prior diffusion-driven phase transformation leads to a decrease of the MS. For the alloying element influence on the transformation start temperatures of the other phases, see Watt et al. (1988). By energetic considerations, Todinov et al. (1996) derived a formulation for the MS as a function of the maximum and minimum principle stresses. In a tensile test, they showed that the MS increases linearly with the applied stress and, in a hydrostatic case, the MS decreases linearly with the pressure. In an ANN framework, Capdvila et al. (2003) showed that, in good agreement with experimental results for various steels, the MS increases with an increase of the austenite grain size which is highly dependent on the heating conditions. Capdvila et al. (2003) state that the "reduction of lattice imperfections and the increase in frozen vacancies owing to higher austenitization temperatures increase martensite nucleation sites and then enhance the transformation". To take this effect into account, Lee and Lee (2008) took the grain size of the steel into account in addition to its chemical composition. Furthermore, Nikravesh et al. (2012) found that

- A reduction of cooling rate can bring about an increase or decrease in MS and martensite formation finish temperature (MF) depending on prior diffusion-driven phase transformation,
- A prior plastic deformation results in a decrease in MS and MF, in particular, at lower cooling rates, and
- The CCR is increased about 40% by applying 40% plastic deformation.

The evolution behavior of the KM model is exponential. However, some steels show an "S"-shape behavior of the volume fraction in dilatation experiments (see, e.g., Miokovic, 2005; Schwenk, 2012). To capture this behavior, Neumann and Böhlke (2016) proposed a sigmoidal function of the temperature dependence of the martensitic volume fraction. It has been shown that this approach leads to a generalization of the KM equation when the model is given in a corresponding rate form. Other heuristic models have been proposed by Yu (1977) and Schröder (1985) taking the MF into account.

Since the KM equation is valid for isothermal phase transformation processes, several authors, see e.g. Ackerström and Oldenburg (2006), Olsson (2009), and Naderi (2007), used this model directly to determine the martensitic volume fraction. This approach is valid, if it is assumed that the martensitic transformation is only temperature-driven and no loading path dependence exists, e.g. prior diffusion-driven transformation or prior deformation. Another method to describe the non-isothermal martensite evolution is to derive an evolution law as described in the following.

Based on an energetic approach considering Gibbs free energy and taking the dissipation equation into account, some authors found generalized KM relations including a stress- and strain-dependent contribution (see, e.g., Fischer et al., 1998; Kubler et al., 2011). Similar to the diffusion-driven phase transformation approach, De Oliveira et al. (2010) suggested a dissipation potential for the martensitic phase transformation leading to a rate law consistent to the usual KM evolution equation. A similar rate law has been proposed by Lee and Lee (2008) taking the austenitic grain size into account.

A totally different approach describing the TRIP effect and the martensitic evolution has been proposed by Ostwald et al. (see, e.g., Ostwald et al., 2010; 2012). They introduced a micro sphere on the micro scale and mapped the macroscopic quantities, e.g. strain, on the discretized directions of the micro sphere. This leads to a one-dimensional formulation along the discretized directions. Via considering Gibbs free energy of the possible phases, the martensitic volume fraction is determined by an energy minimization problem and a probability formulation of the transformation based on the works of Achenbach (1989) and Govindjee and Hall (2000).

The phase transformation has a significant impact on the resulting hardness, e.g., the Vicker's hardness (VH), of a hot stamped part. Each phase has an inherent hardness which can be determined based on the chemical composition of the steel via heuristic equations for hot stamping steels given by Maynier et al. (1978) and modified by Li et al. (1998). For the martensitic phase, Ashby and Easterling (1984) identified a polynomial relation of the hardness to the carbon concentration. The overall Vicker's hardness is given by the mixture law. Ackerström and Oldenburg (2006) proposed their own equation for bainite and martensite and used for ferrite/pearlite the equations proposed by Maynier et al. (1978). The influence of the cooling rate has been taken into account by Bok et al. (2015) by a multiplicative modification of the governing models. Experimentally determined hardness is usually used to validate the simulations. In the context of end-quench tests, see, e.g., Lee et al. (2010) and, in the context of thermomechanical modeling of the hot stamping process, see, e.g., Bok et al. (2015). For the dependence of the hardness on the flow stress, see, e.g., Tabor (1951); Cáceres and Poole (2002).

1.2.5 The Two-Scale Problem - Homogenization and Localization

The hot stamping defining factor is the development of the microstructure during the process. Therefore, the impact of the interaction of the different phases has to be considered in an accurate way. In this context, a distinction is made between three scales:

- The macroscopic (macro) scale, i.e. the scale on which the considerations are of interest,
- The microscopic (micro) scale, i.e. the underlying scale on which the microstructure is defined, and

- The mesoscopic (meso) scale, i.e. the intermediate scale lying in-between the macro and micro scale.

In a lot of applications, e.g., constitutive modeling of polycrystalline materials (see, e.g., Masson and Zaoui, 1999; Jöchen, 2013), a distinction is made between two scales, i.e. the macro and the micro level. If on the micro scale the substructure of a phase is of importance for the overall mechanical behavior, three scales are considered (see, e.g., in the field of SMA, Kouznetsova and Geers, 2008; Mahnken and Wilmanns, 2011). If overall (macroscopic) properties or quantities have to be defined based on the informations on the micro scale, the transition in scales is referred to as homogenization. On the other hand, if the macroscopic quantities have to be redistributed to the different phases on the micro scale, the transition in scales is referred to as localization. A multi-scale simulation method is defined by the following essential requirements (see, e.g., Kanouté et al., 2009):

- Derivation of a localization rule for the redistribution of the macroscopic stress or strain to the different phases on the micro scale,
- Definition of the thermomechanical constitutive behavior of the phases on the micro scale, and
- Definition of the homogenization rule for determination of the macroscopic quantities based on the corresponding microscopic ones.

In this context, three different approaches have emerged in literature (see, e.g., Kanouté et al., 2009):

- The computational methods,
- The analytical and semi-analytical methods, and
- The hybrid methods.

In the field of computational methods, both the finite element (FE) simulations performed at both scales, i.e. the macro and micro scale, referred to as FE^2 methods (see, e.g., Renard, 1987; Feyel, 1999; Miehe et al., 1999; Terada and Kikuchi, 2001; Geers et al., 2010), and the Fast-Fourier-Transformation (FFT) simulation, where on a triangular mesh (see, e.g., Walker KP, 1994; Fotiu and Nemat-Nasser, 1996) or mesh-free approach (see, e.g., Moulinec and Suquet, 1998; Eyre and G.W., 1999; Michel et al., 2000), the strain and stress fields obtained using Fourier series, are of great significance. Due to the spatial resolution of the representative volume element (RVE), the FE^2 method is expensive with regard to the required memory and computational time. For this reason, for the determination of the overall properties, the FFT method gives an alternative approach by solving the Lippmann-Schwinger equation (see, e.g., Kröner and Sauthoff, 1972) based on Fourier series. A memory and computational costs improving FE^2 based approach is the method of statistically similar RVE (SSRVE) shown by Balzani et al. (2014). In the SSRVE method, RVEs with reduced complexity are constructed by minimizing the least square deviation of statistical morphology characteristics between the real and artificial periodic RVE, e.g., the two-point probability or the line-path function (see, e.g., Torquato, 2002). Summarizing, the computational methods direct towards an accurate representation of the microstructural influence on the macroscopic scale. This goes hand-in-hand with substantial numerical costs.

Analytical or semi-analytical methods try to make predictions about the macroscopic behavior based on available statistical information on the microstructure. In contrast to the computational methods, in such methods, the microstructure is not spatially resolved. This leads to numerically efficient methods at the expense of accuracy. In the linear case, Voigt (1889) and Reuss (1929) derived estimates, which coincide with the assumptions of a uniform strain or uniform stress field on the microstructure. Hill (1952) showed that the method of Voigt (1889) is an upper bound for the overall stiffness tensor. In the same way, the method of Reuss (1929) leads to a lower bound for the stiffness. Later, Hashin and Shtrikman (1962a) extended the method of minimization of potential energies by introducing a homogeneous comparison material (CM). For infinite large magnitude and a magnitude of zero of the stiffness of the CM both the upper and the lower bound, respectively, are included in the Hashin-Shtrikman (HS) method. Furthermore, tighter bounds for the effective stiffness of the composite can be found in the HS method by choosing the stiffness of the CM in an appropriate way. For a microstructure consisting of a distinct phase, referred to as matrix, and an inclusion, by choosing the stiffness of the CM to be the overall stiffness of the composite, the HS method contains the self-consistent (SC) method, which was originally proposed by Kröner (1958) (see also, e.g., Hill, 1965a; Budiansky, 1965). Since the overall stiffness is not known a priori, the SC method is an implicit estimate.

In the nonlinear case, as a first approach, the simple assumption of a uniform stress or strain state on the microstructure has been proposes by Sachs (1928) and Taylor (1938), respectively. The latter has been extended by Lin (1957) for uniform strain rates on the microstructure. Considering the minimum of the overall potential energy and its complementary, the Taylor and Sachs estimates appear as an upper and lower bound, respectively, for the overall mechanical behavior. Based on a linearization of the constitutive behavior of the phases, Hill (1965b) suggested to apply the SC method to the linearized model in an incremental way. Since the method was derived for elasto-plastic materials, Hutchinson (1976) extended the method for visco-plastic phases. Several authors realized that the linearization of the constitutive law is not unique. Due to the ambiguity of the linearization, three further methods have been proposed by Berveiller and Zaoui (1979) using the secant linearization, Molinari et al. (1987) using a tangent linearization, and Masson et al. (2000) using an affine formulation. On the other hand, Willis (1983) generalized the variational method of HS for nonlinear phase behavior, which led to the bounding technique of Talbot and Willis (1992) by introducing nonlinearities to the homogeneous CM (see, e.g., Jöchen,

2013). Based on this bounding technique, Jöchen and Böhlke (2012) suggested a set of estimates for crystal-plastic phases on the micro scale by using the stiffness of the CM as a further parameter. In their work, Jöchen and Böhlke (2012) showed the influence of this parameter on the evolution of the grain orientation in a deep-drawing process. In the same way as in the linear case, the first-order bounds of Sachs and Taylor are included in the nonlinear HS approach by setting the magnitude of the stiffness of the CM as zero and infinity, respectively. In the context of polycrystalline microstructures, Ponte Castañeda (1992) used a heterogeneous linear CM to derive a class of bounds which contain the Talbot-Willis bounds. Later, Ponte Castañeda and Suquet (1998) showed that in a modified secant frame, the higher-order theory based on second-order moments in each phase of the linear CM proposed by Suquet (1995) corresponds to the method used by Ponte Castañeda (1992). Since Leroy and Ponte Castañeda (2001) found that, in some cases, the second-order estimates can violate the HS bounds, Ponte Castañeda (2002) improved the second-order approach by incorporating field fluctuations.

Hybrid methods try to combine the advantages of the computational and semi-analytical methods. For this purpose, Dvorak (1992) introduced a method, referred to as transformation field analysis (TFA), which takes the resolved microstructure in a preliminary step into account. The phases on the microstructure are divided into subvolumes with piecewise uniform material properties and eigenstrains, e.g., thermal or plastic strains. Solving linear problems via FEM leads to the localization operations for the strain and the eigenstrains. As special cases, several homogenization methods, e.g., Taylor, Sachs, or SC, are included in the TFA method. Since the TFA method yields macroscopic stress responses which are too stiff (see, e.g., Kanouté et al., 2009), Chaboche et al. (2001) introduced a corrected TFA method based on an asymptotic tangent stiffness tensor. However, several authors (see, e.g., Teply and Dvorak,

1988; Chaboche et al., 2001) realized that too many subvolumes are needed to obtain acceptable macroscopic results. To overcome that problem, Michel and P. (2003) and Michel and P. (2004) introduced the non-uniform TFA (NTFA) method by introducing non-uniform transformation fields, referred to as modes, to approximate the inelastic strains on the microstructure. For further discussion, see, e.g., Fritzen (2011) and Kanouté et al. (2009).

In the context of hot stamping, a simple mixture law is usually used to take the microstructure into account. However, in the field of modeling the TRIP effect, homogenization techniques are used by several authors. For example, Leblond et al. (1986) used a Sachs, Fischer et al. (1992) a Taylor-Lin, and Diani et al. (1993) and Patoor et al. (1988) an SC approach to derive their TRIP relations. Since, in most thermomechanical approaches to multiphase steels, an extended Leblond relation for the GJ effect is used, the modeling results in a combination of a mixture law and a Sachs homogenization (see, e.g., Ackerström et al., 2007; Wolff et al., 2008b; De Oliveira et al., 2010; Bok et al., 2014).

1.2.6 Modeling of Finite Deformations in the Hot Stamping Context

Since most models used for the simulation of the hot stamping process are derived in a small deformation framework (see, e.g., Ackerström et al., 2007; Wolff et al., 2008b; De Oliveira et al., 2010; Bok et al., 2014), the extension to finite deformations, which take place during the forming stage by finite rotations, as well as finite deformations, is done by use of the updated Lagrangian formulation (ULF) (see, e.g., Liu et al., 2006). In contrast to the total Lagrangian Formulation (TLF), in the ULF, the reference equilibrium placement is the placement at the current time. A geometrical linearization of the highly nonlinear formulation of the virtual work between the current time and the incremented time makes it possible to use the small deformation models in an incremental form for the simulation of the finite deformation process (see, e.g., Zieliński and Frey, 2001; Wriggers, 2008).

In the context of modeling thermomechanical processes including phase transformation, the TLF approach has been used by, e.g., Hallberg et al. (2010), and Mahnken et al. (2012) by a modified finite deformation formulation and a multi-phase extension of the model, which was originally proposed by Hallberg et al. (2007). In this thermodynamical approaches, the TRIP effect has been taken into account by an extension of Leblond's model, the phase transformation by a thermodynamically consistent consideration and the multiphase behavior by a simple mixture rule. Böck and Holzapfel (2002) found an interpretation of Leblond's formula as a viscoplastic evolution law of Perzyna-type (Perzyna, 1971) with a von Mises yield function and zero yield stress. With this interpretation, Böck and Holzapfel (2002) were able to follow the extension of the von Mises plasticity for large deformation given by, e.g., Simo and Hughes (1998) to find an expression for the finite strain form of Leblond's equation. Furthermore, Ostwald (2015) extended his two-scale microsphere model to a finite deformation formulation in a thermodynamic framework.

1.2.7 Final Remarks on the Modeling of Hot Stamping

With the models suggested by, e.g., Bok et al. (2014), Ackerström et al. (2007), Hochholdinger (2012), some of the challenges in hot stamping defined by Neugebauer et al. (2012) have been taken up. To realize an optimized cycle time, cooling systems with regard to shape, number, and position of the cooling channels (see, e.g., Michelitsch and Mehnen, 2006; Steinbeiss et al., 2007; Liu et al., 2013) and cooling systems with regard to cooling medium and flow rate of the cooling medium (see, e.g., Aziz and Aqida, 2013) have been investigated. In the context of process' safety, Hu et al. (2015) suggested a model for press hardenable steels based on the damage model by Lemaitre (1985). In FE simulations, Hu et al. (2015) considered the influence of the blank holding force and friction on the damage evolution as well as the location of potential crack initiation which has been in good agreement with the experimental observation. Further works in this fields include Dahan et al. (2006), Lin et al. (2013), Shi et al. (2015).

1.3 The Thermo-Micromechanical Model and Structure of the Thesis

With regard to hot stamping, a constitutive model has to be able to

- Capture the complex phase decomposition of austenite into ferrite, pearlite, bainite, and martensite,
- Consider the thermo-elasto-plastic behavior of the bulk,
- Render the non-classical plastic effect TRIP, in particular, the GJ effect, and
- To describe the nonlinear thermomechanical two-scale problem.

In literature, the models which have thus far been proposed in the context of hot stamping or modeling of thermomechanical processing of steel, focus on the GJ effect via extensions of Leblond's formula and on the correct rendering of the phase transformation via modifying the JMAK or KV models. The plasticity is taken into account by an effective phenomenological modeling with one effective internal variable for all phases. The two-scale problem is considered by simple mixture rules for the elastic, thermal, and yield properties of the bulk (see, e.g., Ackerström et al., 2007; Wolff et al., 2008b; De Oliveira et al., 2010; Hochholdinger, 2012; Bok et al., 2014).

In contrast to this phenomenological constitutive modeling, a more physically based two-scale approach is suggested based on a

- Thermodynamical consistent thermo-elasto-plastic model for each distinct phase via an introduction of a Helmholtz free energy for each phase,
- Nonlinear isotropic hardening of each phase by a temperature-dependent Voce-type or Swift-type law,
- Modified JMAK or KV model taking the cooling rate into account for the correct description of the diffusion-driven phase transformation,
- Nonlinear extended KM rate law capable of capturing the "S"-shape behavior of the martensitic phase transformation considered for some steels,
- Leblond-type evolution law for the rendering of the GJ effect of each growing phase in the parent phase austenite, and
- Thermomechanically extended HS-type homogenization scheme for the scale transition originally suggested by Jöchen and Böhlke (2012).

The starting point for the constitutive modeling are experimental observations made in both thermomechanical tensile experiments and experiments with the hot stamping of a w- and u-shaped parts described in Karbasian (2010). In Chapter 2, the stress-strain-curves, resulting shapes of the hot stamped parts, and residual stress measurements are shown.

After a short composition of the governing continuum mechanical results in Chapter 3, the thermo-elasto-plastic behavior of a single phase is derived by the introduction of a Helmholtz free energy and use of the Clausius-Duhem (CD) inequality. Furthermore, models for the transformation and TRIP strain are suggested. The nonlinear isotropic hardening of the phases in the steel grade 42CrMo4 and 22MnB5 is modeled via a Voce- and Swift-type law, respectively. The phases' yield curves of 42CrMo4 show a highly nonlinear hardening until a strain of about 0.01 and a linear hardening beyond this transition region well captured by the Voce-type hardening. The phases' Swift hardening parameters of 22MnB5 are given by Bok et al. (2014) and are used directly in this work. Since viscous effects are neglected, the Swift law captures also used the hardening of the austenitic phase. This chapter is concluded by considering the phase transformation determining evolution equations.

The localization and the homogenization for the mechanical and thermal problem is discussed in Chapter 5. For the distribution of the overall strain to the phases, the HS-type estimate is introduced taking the inelastic and thermal strains as eigenstrains into account. The determination of the overall thermal conductivity, the second-order homogenization scheme of HS is presented. In addition, the special form of the heat equation determined by the constitutive model and aspects of the thermodynamical consistency are discussed.

Chapter 6 enlightens the numerical implementation of the constitutive model into the commercial FE simulation tool ABAQUS (2016). For this purpose, the user interfaces UMAT, UMATHT, and UEXPAN of ABAQUS (2016) are used for the mechanical, thermal, and thermal expansion problem, respectively.

At an early stage of the work, the material parameters of the steel grade 22MnB5 were not available. To be able to investigate the influence of the latent heat and transformation strain in thermomechanical processes, the steel 42CrMo4 was considered. For the steel grade 42CrMo4, the entire set of material parameters can be found in the literature. For this reason, the parameters of the model and the process parameters are identified based on the works of Tang et al. (2014) and Bok et al. (2014) for 22MnB5 and Miokovic (2005) and Schwenk (2012) for 42CrMo4.

To be able to use the homogenization schemes mentioned above, viscous effects are neglected and static flow curves are used as an first approximation. Since the forming step in hot stamping is performed with high stamp velocities, the parameters of the hardening law are identified based on the strain-stress curves gained by the fastest forming.

The process parameters are identified by the group of Merklein, see, e.g., Merklein et al. (2009), Lechler et al. (2010), or Geiger et al. (2005). The parameter identification is shown in Chapter 7.

With the suggested thermomechanical two-scale model both the influence of the latent heat and transformation strain in thermomechanical processes is investigated using the steel grade 42CrMo4 and the experimentally investigated hot stamping of 22MnB5 sheet metals is simulated. The numerical results and the comparison with the experimental observations are shown in Chapter 8.

The summary and outline in Chapter 9 and the Appendix conclude this work, where linear homogenization methods, further material parameters from literature, and further numerical results are discussed.

Chapter 2

Experimental Investigation of the Hot Stamping Process

2.1 The Experimental Setup of the Hot Stamping Process and Results

Experimental investigation into the hot stamping process was performed at the Institute of Forming Technology and Lightweight Construction (IUL) in Dortmund in cooperation with the Fraunhofer Institute for Mechanics of Materials IWM in Freiburg. The steel used for the experiments was provided by the company voestalpine (2016) and has the chemical composition given in Table 2.1. The Zn-Ni coated phs-ultraform 1500 Z140 (2016) sheet metals with a thickness of 1.8mm were originally developed for the indirect hot stamping process (see, e.g., phs-ultraform 1500 Z140, 2016; Steinhoff et al., 2012) and have also been used in the direct hot stamping process recently (see, e.g., Materials Views, 2016; phs-ultraform, 2016). Since vaporized Zinc oxide is hazardous to health, voestalpine (2016) advises that the sheet metal must not to be heated beyond 1203.15K. According to the company's statement, a totally austenitized microstructure in the sheet metal is ensured when austenitization takes place at a temperature of 1143.15K for 45s. Longer austenitization times have an influence only on the corrosion protection of the layer. Furthermore, for comparison purposes, the phase transformation behavior of the phs-ultraform 1500 Z140 (2016) in the continuous cooling case is compared with a similar steel of the same grade (see Table 2.1) developed by thyssenkrupp (2016) and investigated by Naderi (2007).

Table 2.1: Chemical composition in weight percent of the steel grade 22MnB5 of voestalpine (2016) phs-ultraform 1500 Z140 (2016) and of thyssenkrupp (2016) MBW 1500 (2016)

	voestalpine (2016)	Faderl and Radlmayr (2006)	BMW AG (2009)	Naderi (2007)
C_C	0.20-0.25	0.22	0.224	0.23
C_{Si}	0.5	0.2	0.227	0.22
C_{Mn}	2.0	1.2	1.213	1.18
C_P	0.02	-	0.013	-
C_S	0.005	-	<0.001	-
C_{Cr}	0.5	0.25	0.231	0.16
C_{Al}	0.02-0.1	-	0.047	0.03
C_{Ti}	0.05	-	0.033	0.04
C_B	0.002-0.005	0.003	0.003	0.002

For the investigation of the hot stamping process two different demonstrators are considered. The tools for the manufacturing of the demonstrators have been developed by the DFG Forschergruppe 552 (2016) and are depicted in Figure 2.1. Based on the sheet metal's initial geometry with a length of 430mm and width of 190mm, the hot stamping process is performed with and without a counter punch to manufacture the w-shape and the u-shape geometries, respectively. For these purposes, the direct hot stamping procedure with the experimental set up depicted in Figure 2.2 is performed. The set up consists of a chamber furnace of the type EGH 40/30/65 Fabr. Allino, a single column drawing press of the type HPSZK 100-1025/650 (for details see Appendix A), and measuring devices as a thermal imaging camera AGEMA 570 produced by the company Flir Systems, Inc. (2016), thermal sensors 1mm beneath the surface of the forming tools, and a displacement transducer to capture the punch displacement (see Figure 2.4).

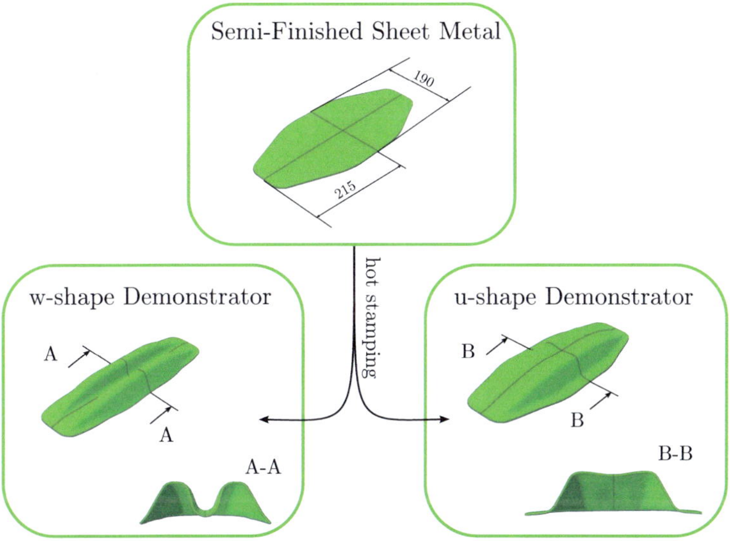

Figure 2.1: The w-shape and the u-shape demonstrator investigated in hot stamping

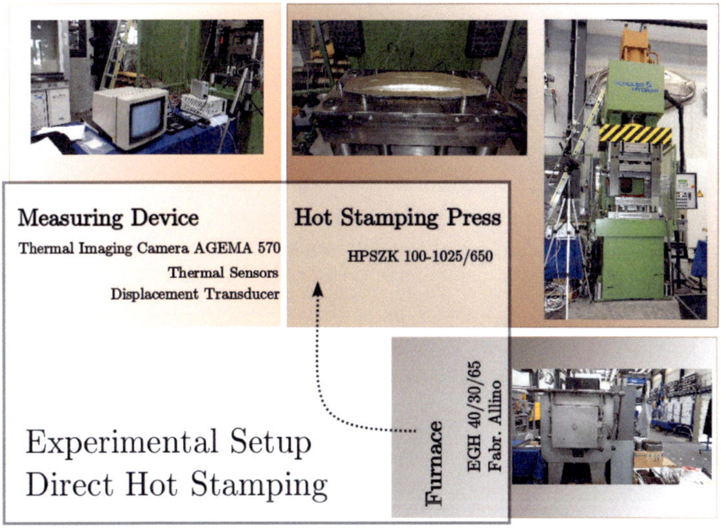

Figure 2.2: Experimental setup of the hot stamping experiments

The hot stamping process is performed in three steps:

1. Austenitization of the sheet metal in the furnace,
2. Manual transfer of the austenitized sheet, and
3. Forming and quenching in the closed deep drawing press depicted in Figure 2.5.

Two parameters define the austenitization of the sheet metal: the austenitization time and austenitization temperature. Thereby, the time of the sheet metal in the furnace comprises of both the time to reach the austenitization temperature after insertion, referred to as temperature set-up time, and the time at a constant temperature level, referred to as austenitization time. The temperature set-up time is determined by

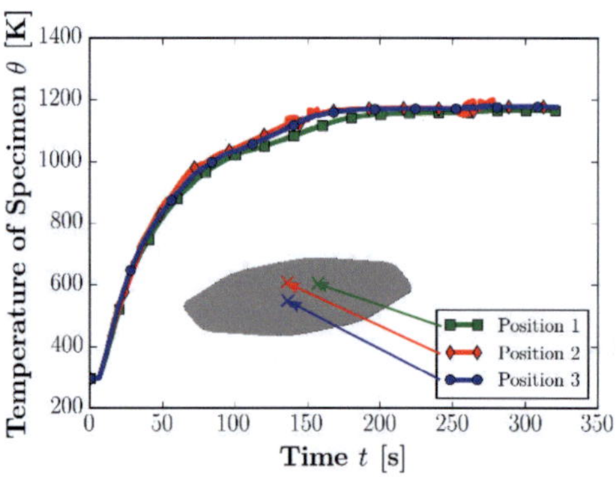

Figure 2.3: The determination of the temperature set-up time

heating experiments in the furnace with temperature sensors on the sheet metal (see Figure 2.3). The course of the temperature curve indicates that at a temperature of about 1000K, the energy consuming phase transformation to austenite takes place. A constant temperature level is reached at about 265s. In accordance with the statement, the temperature window for austenitization is above 1143.15K and below 1203.15K with a minimum austenitization time of 45s. For the investigations in this work, two austenitization temperatures, i.e., 1150.15K and 1173.15K, and two austenitization times, i.e., 45s and 180s, are considered. This two austenitization times are far beyond the time resulting from the heuristic formula by Lechler (2009)

$$t_\gamma^{min} = 1750.6\text{s}\exp\left(-0.0067\frac{1}{^\circ\text{C}}\theta_\gamma + 0.3822\frac{1}{\text{mm}}\left(1.75\text{mm} - s_0\right)\right), \tag{2.1}$$

where θ_γ is the austenitization temperature in °C and s_0 the sheet metal thickness in mm.

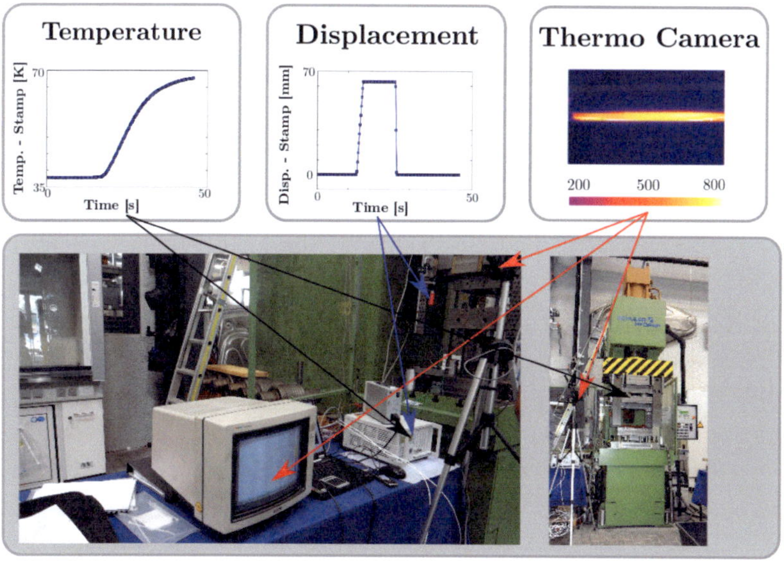

Figure 2.4: Experimental matrix of the hot stamping experiments

The transfer of the austenitized sheet metal is performed manually. Thereby, it has to be ensured that the forming takes place in an austenitic state of the sheet metal. Lenze and Sikora (2006) have given guide values for the cooling rate of an austenitized sheet metal in air, i.e., 22K/s for a sheet metal thickness of 1.5mm and 12K/s for a sheet metal thickness of 2.5mm. Linear interpolation yields a cooling rate of 19K/s for a sheet metal thickness of 1.8mm, which is considered in this work. The ferritic phase transformation takes place at a temperature of 1042.24K which leads to a maximum transfer time of 6.89s. In the experiment, careful attention was paid to realize a transfer time of 6s.

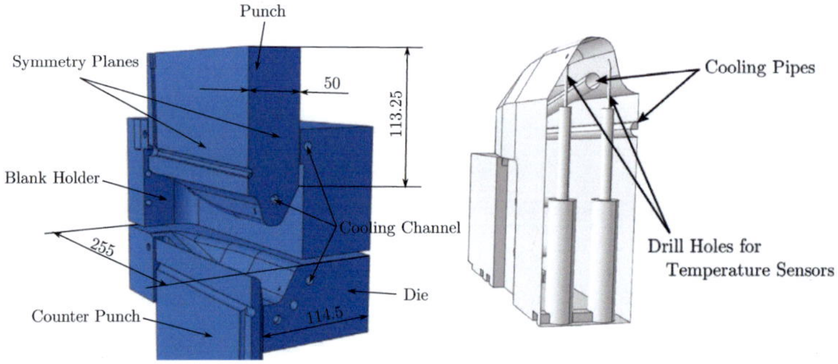

Figure 2.5: Quarter of the deep drawing press (left) and interior of the punch with drill holes for the temperature sensors (right)

The investigation carried out by Karbasian (2010) showed that the hold time influences both the microstructure and the resulting hardness of hot stamped parts manufactured in the direct as well as in the indirect process. As a result, the forming time and the hold time on both the shape and residual stress development are considered in the experimental investigations. Thereby, deformation rates of 100% and 50% of the maximum stamp velocity of 350mm/s, and hold times of 6s, 8s, and 10s are adjusted.

To provide the possibility of cooling the forming tools to ensure standardized initial conditions for the process, cooling pipes have been taken into account in the construction of the forming tools (see, e.g., Steinbeiss et al., 2007; Karbasian, 2010). However, a fluid circulation system was not available for this feature to be used. Thus, to ensure the cooling of the forming tools to the standardized condition, only ten forming experiments per hour were able to be performed.

In summary, the influence of the process variables austenitization time, austenitization temperature, deformation rate, and hold time on the final shape and the residual stress development are considered with the values given in Table 2.2. The notation of the parameter set used in a hot stamping process is listed in the following way:

aust. temp. - aust. time - deformation rate - hold time,

e.g., 880-45-100-6 for a process with an austenitization temperature of 880°C, an austenitization time of 45s, a deformation rate of 100%, and a hold time of 6s. For statistical certainty, the hot stamping process is performed three times for each parameter set. The repetition number is only listed in cases in which it is necessary.

Table 2.2: Investigated process variables

Process variables	Investigated values
Austenitization temperature	880°C and 900°C
Austenitization time	45s and 180s
Deformation rate	50% and 100% of 350mm/s
Holding time	6s, 8s, and 10s

2.2 Shape of the Hot Stamped Parts and Residual Stresses

To compare experimental results of different process parameter sets and to validate numerical results, the final shape of the hot stamped parts and the residual stresses measurements are used. The geometry measurements have been performed by the IUL in Dortmund using the industrial

3D scanning technology Atos system produced by the company GOM GmbH (2016). The shape deviation of the parts manufactured with different parameter sets is compared for the entire field as depicted in the bottom of Figure 2.6 for the w-shape configuration with the process parameter sets 880-180-50-6 and 880-180-100-10. Furthermore, the cross sections are considered in detail at the positions A-A and B-B as depicted in the top of Figure 2.6 with the resulting curves given in Figure 2.7.

In this work, the deviations of curves gained in experiments and numerical computations are measured with the L^2-norm-induced metrics. For example, the deviation of the functions $f(x)$ and $g(x)$ is determined by

$$d_{L^2}\left(f(x), g(x)\right) = \|f(x) - g(x)\|_{L^2} = \sqrt{\int_{x_s}^{x_e} \left(f(x) - g(x)\right)^2 \, \mathrm{d}x}. \quad (2.2)$$

Equivalently, based on the L^2-norm-induced metrics, the relative deviation of the functions is defined by

$$d_{L^2}^{rel}\left(f(x), g(x)\right) = \frac{\|f(x) - g(x)\|_{L^2}}{\|f(x)\|_{L^2}} = \frac{\sqrt{\int_{x_s}^{x_e} \left(f(x) - g(x)\right)^2 \, \mathrm{d}x}}{\sqrt{\int_{x_s}^{x_e} f(x)^2 \, \mathrm{d}x}}. \quad (2.3)$$

Note that, in contrast to the metrics d_{L^2}, the relative deviation $d_{L^2}^{rel}$ is not symmetric and, due to this, no metrics. Furthermore, if $f(x) = 0$ holds, the relative deviation is not defined and the roles of $f(x)$ and $g(x)$ have to be interchanged. Nevertheless, since the relative deviation is positive definite, fulfills the triangle inequality, and is quasi symmetric, i.e. the difference between $d_{L^2}^{rel}\left(f(x), g(x)\right)$ and $d_{L^2}^{rel}\left(g(x), f(x)\right)$ is slight if $f(x) \approx g(x)$ holds, the relative deviation is suitable to get an idea of how much the curves deviate from one another.

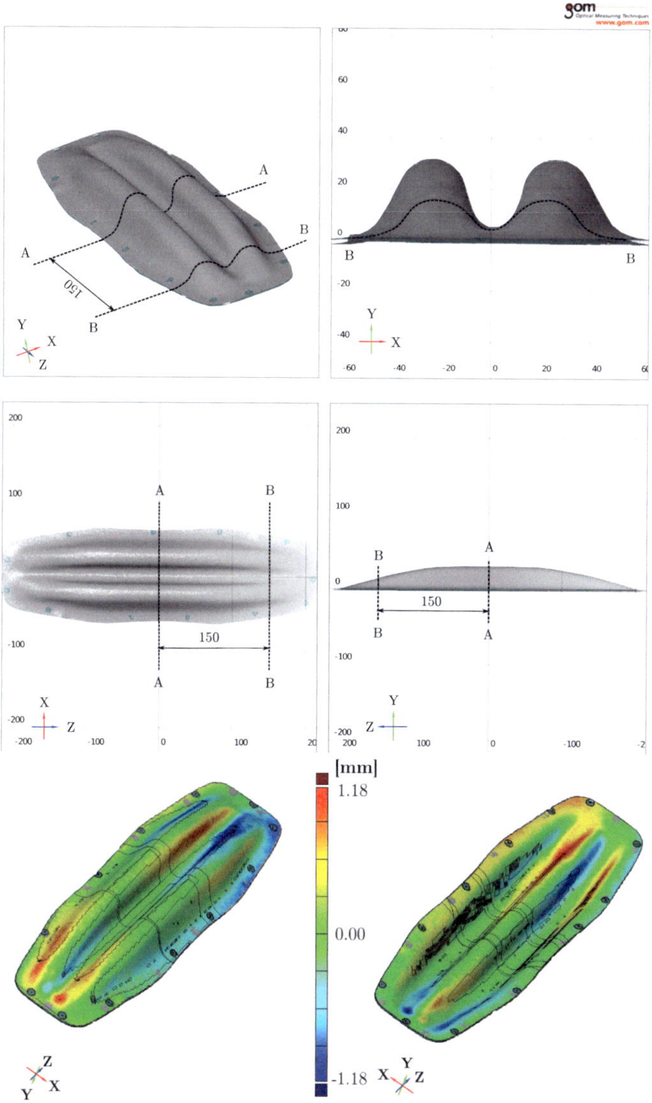

Figure 2.6: Hot stamped w-shaped part of the process parameter set 880-180-100-10 (top and middle) and comparison of the final w-shape of the process parameter set 880-180-50-6 and 880-180-100-10 (bottom)

In Figure 2.7, the cross sections at the positions A-A and B-B of the w-shape configuration with an austenitization temperature of 880°C and austenitization time of 180s are depicted. The deformation rate and the hold time are variated between 50% and 100%, and 8s and 10s. The total deviation between the resulting shape of the parameter sets $d^t_{L^2}$ is determined via averaging between the deviation between the upper curves $d^{\text{A}-A,upper}_{L^2}$ and lower curves $d^{\text{A}-A,lower}_{L^2}$ at the position A-A, and the upper curves $d^{\text{B}-B,upper}_{L^2}$ and lower curves $d^{\text{B}-B,lower}_{L^2}$ at the position B-B

$$d^t_{L^2} = \frac{d^{\text{A}-A,upper}_{L^2} + d^{\text{A}-A,lower}_{L^2} + d^{\text{B}-B,upper}_{L^2} + d^{\text{B}-B,lower}_{L^2}}{4}. \tag{2.4}$$

Equivalently, the total relative deviation between the parameter sets $d^{\text{rel},t}_{L^2}$ is defined. For the variation of the deformation rate and the hold time, the total relative deviations read

$$\begin{aligned} d^{\text{rel},t}_{L^2}\left(y_{\ldots\text{-50-8}}, y_{\ldots\text{-100-8}}\right) &= 0.049994, \\ d^{\text{rel},t}_{L^2}\left(y_{\ldots\text{-50-8}}, y_{\ldots\text{-100-10}}\right) &= 0.061165, \\ d^{\text{rel},t}_{L^2}\left(y_{\ldots\text{-100-8}}, y_{\ldots\text{-100-10}}\right) &= 0.101093. \end{aligned}$$

For the considered parameter sets, the deviation field depicted in the bottom of Figure 2.6 has a maximum deviation of ± 1.18mm and maximum relative deviation of about 3.5% if related to the maximum displacement of about 33mm. Since the influence of the process parameters on the relative deviations of the cross sections and the field comparisons are quite low with up to 10%, the residual stresses and microstructural composition of the hot stamped parts are investigated. For the determination of the residual stresses in the finished parts, the hole-drilling method (see, e.g., Rendler and Vigness, 1966; Schajer, 1988; Karbasian, 2010) is used. It is also desirable to have the texture information of the point where the hole is drilled. Since the hole-drill method is destructive and for texture analysis an area has to be cut out of the hot stamped part, the

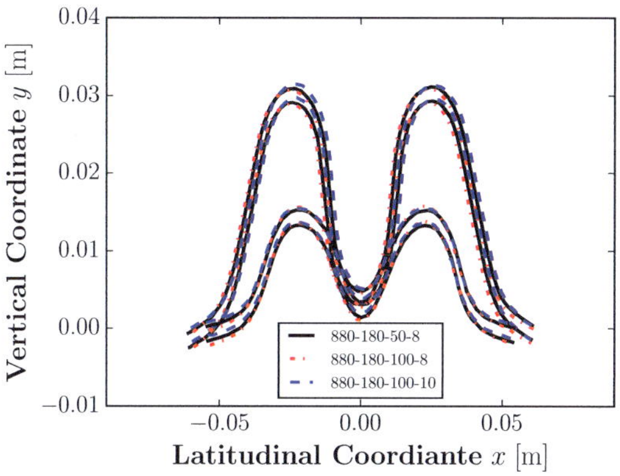

Figure 2.7: Comparison of the final w-shape at the cut A-A and B-B of the process parameter sets 880-180-50-6, 880-180-100-8 and 880-180-100-10

best approach is to choose this point for texture measurements which is symmetrical in longitudinal direction (LD) to the point where the hole is drilled. As depicted in Figure 2.8, the residual stresses and the texture measurements are performed on the symmetry plane in transversal direction (TD) at the highest point of the w-shaped cross section.

Considering Figure 2.9, which shows the residual stresses in LD and TD, one can observe that residual stresses up to 300MPa are present in the hot stamped part and that the residual stresses decrease approximately exponentially with depth. While the decreasing tendency of the residual stress progression in the sheet metal thickness is the same for all considered process parameter sets, it is not possible to make clear statements regarding the influence of the process parameters. Even for the two repetitions of the same parameter set 880-180-100-8, both the residual stress curve and the residual stress level are different. The microstructure analysis of these two parts (Figure 2.8), detects that the amount of the high strength phases, martensite and bainite, is higher in the second specimen while the first specimen contains a huge amount of ferrite. This leads to the assumption, that, even for equal process parameter sets, small variations of the tool temperature or the manual transportation time have a huge impact on the thermal treatment, due to this, on the microstructural evolution, and, finally, on the residual stresses in the final part.

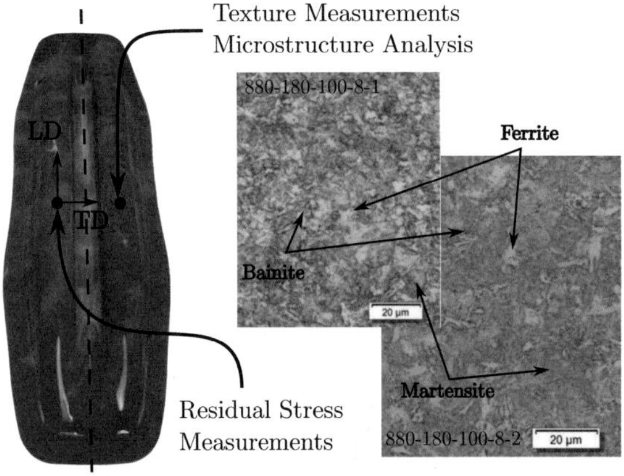

Figure 2.8: Residual stress measurements and microstructure analysis in a hot stamped part (Neumann et al., 2017)

However, two observations can be made. The measurements suggest that

- An increase of the hold time leads to a decrease of the residual stress level, in particular, inside of the sheet metal and
- A decrease of the austenitization time leads to a buckle in the residual stress curve and a higher residual stress level inside of the sheet metal.

A possible explanation for the first observation could be, that the cooling process is not finished for shorter hold times. Thus, ferrite occurs in the microstructure leading to higher residual stresses due to tension between the martensitic and ferritic volumes. On the other hand, the second observation could be explained by the fact that for shorter austenitization times the sheet metal is not entirely austenitized and ferrite remains inside the sheet metal. While the volumes close to the surface are mainly martensite after quenching, inside the sheet metal, the ferrite is still present leading to tension between the martensitic and ferritic volumes and a buckle in the curve.

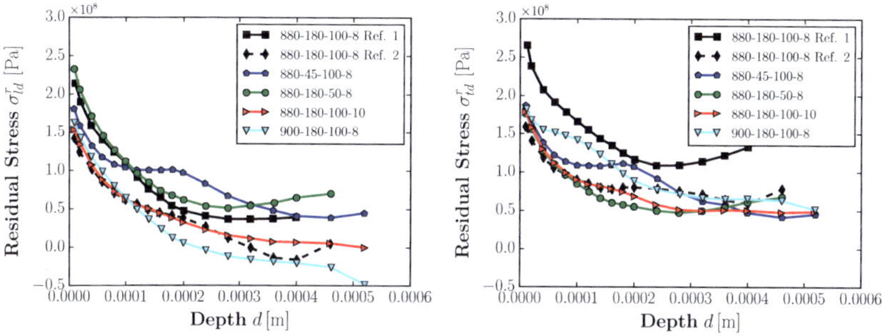

Figure 2.9: Residual stress in longitudinal and transversal direction of a hot stamped part (Neumann et al., 2017)

2.3 Mechanical Material Properties of Hot Stamped Parts

The mechanical behavior of thermomechanically treated sheet metals is investigated to get references for the modeling approach. Therefore, tensile tests of differently thermomechanically loaded specimens are performed and compared to one another. The tensile tests are performed in accordance with the DIN EN 10002 (see, e.g, Klein, 2007) at an universal testing machine Z250/SN5 of the company Zwick Roell (2016). Objects of investigation are

- Anisotropy of the rolled sheet metal in the supplied condition,

- Anisotropy of austenitized specimens cooled and formed in different ways,
- Influence of the austenitization temperature, austenitization time, deformation velocity, and hold time on the mechanical behavior of hot stamped parts.

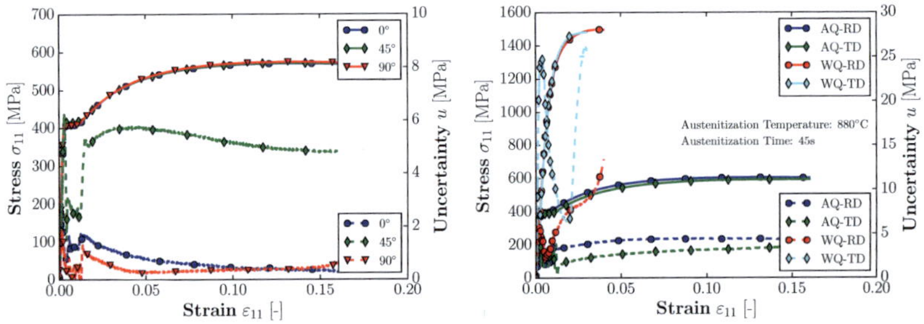

Figure 2.10: The anisotropic stress-strain behavior of the specimens in initial condition (left) and heat treated condition (right) (solid lines: stress and doted lines: uncertainty)

In a first step, the anisotropy is considered. For this purpose, flat tensile specimens in supplied condition and heat treated condition (without forming) are cut out with a laser cutter in rolling direction (RD) and transversal direction (TD), and tensile tests are performed. Heat treated specimens are first austenitized at different temperatures for 180s and second, air-cooled or water-quenched to room temperature. In a second step, the forming influence is considered. For this purpose, tensile specimens are cut out of the flat flanks of the w-shaped hot stamped parts and tensile tests are performed.

The deviations of the resulting stress-strain curves are measured with the L^2-norm-induced metrics and the corresponding relative deviation given in equation (2.2) and (2.3), respectively. Additionally, since the experiments for the same thermomechanical load are performed three times, the uncertainty is considered as well. For n repeated measure-

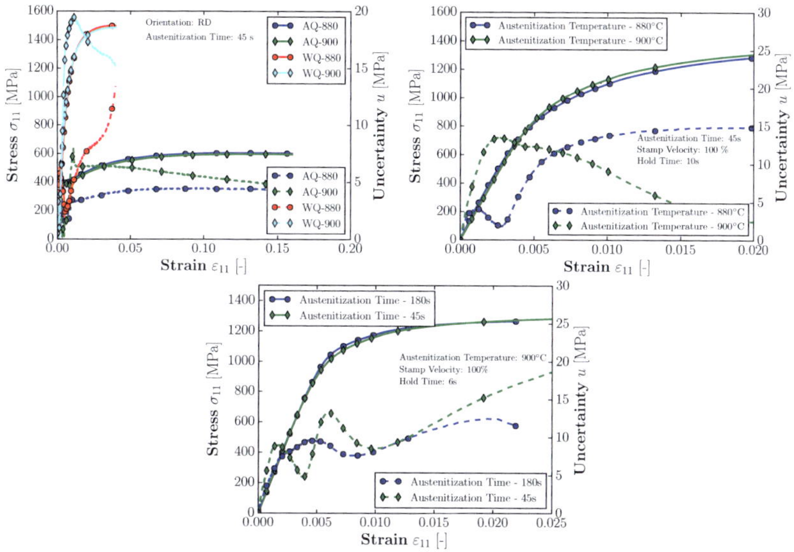

Figure 2.11: Influence of the austenitization temperature and the austenitization time on the mechanical properties of the final part (solid lines: stress and doted lines: uncertainty)

ments of each thermomechanical load, the uncertainty is given by

$$u = \sqrt{\frac{1}{n}\sum_{i=1}^{n}\left(\sigma_{11,i} - \tilde{\sigma}_{11}\right)^2}, \tag{2.5}$$

where $\tilde{\sigma}_{11}$ is the mean value of the stresses in 11 direction σ_{11}, which depend on the strain in 11 direction ε_{11},

$$\tilde{\sigma}_{11} = \frac{1}{n}\sum_{i=0}^{n}\sigma_{11,i}. \tag{2.6}$$

First of all, the influence of the cooling rate is shown on the right-hand side of Figure 2.10. While water-quenched specimens have a tensile strength of about 1500MPa, the tensile strength of the air quenched parts is about 600MPa. In the first case, due to the high strength, one can conclude that the microstructure is almost martensitic. The critical cooling rate (CCR), i.e., the minimum cooling rate which is needed to realize a martensitic microstructure, is exceeded. Air quenching yields a cooling rate below the CCR and a microstructure consisting of ferrite and pearlite.

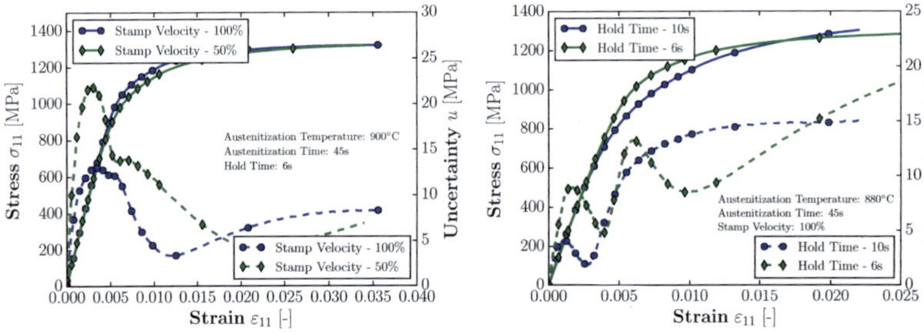

Figure 2.12: Influence of the stamp velocity and of the hold time in the closed tool on the mechanical properties of the final part (solid lines: stress and doted lines: uncertainty)

Furthermore, Figure 2.10 shows the mechanical anisotropy of the material in initial state on the left- and of the material after heat treatment on the right-hand side. In both cases, the anisotropy makes a difference of below 3% in the mechanical tensile behavior (see also Table 2.3). Thus, it can be concluded that the mechanical isotropy of the phases austenite, ferrite, pearlite, bainite, and martensite is a satisfactory assumption for the modeling approach.

The influence of the austenitization parameters on the resulting mechanical behavior is shown in Figure 2.11 for both cases, the unformed heat treatment (top left) and the hot stamping process (top right and bottom). Neither in the unformed case nor in the hot stamping case, the austenitization parameters, i.e. temperature and time, play a significant role, since the maximum relative deviation of the resulting stress-strain curves due to variation of the parameters is below 3% (see also Table 2.3).

The influence of the austenitization parameters on the austenitization process of the specimen in initial condition has not to be taken into account. However, since the maximum resulting tensile strength is about 1300MPa which deviates from the tensile strength of martensite of about 1500MPa, in hot stamping, the mechanical behavior of the final part indicates that the resulting microstructure consists of martensite and weaker phases as bainite, ferrite, and pearlite.

Table 2.3: Influence of the anisotropy, the austenitization process, and the process parameters on the mechanical properties of the initial and final part by use of the L^2-norm-induced metrics d_{L^2} and $d_{L^2}^{rel}$

Influence of the ...	Total Dev. d_{L^2} [MPa]	Rel. Dev. $d_{L^2}^{rel}$ [-]
aniso. - initial part - RD to 45°	1.15	0.0054
aniso. - initial part - RD to TD	1.89	0.0089
aniso. - initial part - 45° to TD	1.41	0.0066
aniso. - air cooled - unformed - RD to TD	6.75	0.03
aniso. - water quench. - unformed - RD to TD	3.09	0.0118
aust. temp. - air cooled - 880° to 900°	4.06	0.018
aust. temp. - water quench. - 880° to 900°	5.86	0.0224
aust. temp. - formed - 880° to 900°	3.89	0.0259
aust. time - formed - 180s to 45s	2.0	0.0124
stamp velocity - 100% to 50%	4.96	0.022
hold time - 10s to 6s	7.3	0.0469

During hot stamping, the austenitization parameter and the stamp velocity, i.e. the deformation rate, causes slight variations in the final mechanical behavior with a relative deviation of about 2.2% (see right-hand side of Figure 2.12). In contrast, the hold time, which corresponds to the cooling time, shows the greatest impact on the mechanical behavior. Even if the relative deviation of the resulting stress-strain curves of the different hold times is only about 4.7%, the yield stress, the tensile strength, and the curve shape differ significantly from one another (see left-hand side of Figure 2.12). The different cooling times lead to an uncompleted decomposition of the austenitic phase which is completed due to air cooling after removal from the press.

In summary, it is concluded that

- The initial mechanical state of the material is mechanically isotropic,
- After quenching without external mechanical load, in both cases the low cooling with air and quenching with water, the mechanical behavior is isotropic,
- Due to this, the austenitic, the ferritic, and the martensitic phases can be assumed to be mechanically isotropic,
- The influence of the austenitization temperature and austenitization time on the mechanical behavior is negligible,
- In hot stamping, the microstructure consists of martensite, bainite, ferrite, and pearlite,
- During the hot stamping process, the hold time in the closed press, i.e. the quenching time, has the greatest influence on the resulting mechanical behavior.

Chapter 3

Basics of Continuum Mechanics

The objective of continuum mechanics is the description of the deformation of continuous materials under thermomechanical influences from the environment. Thereby, on the one hand, statements concerning both the kinematics and the balances of mechanical and thermodynamical quantities, i.e. the linear momentum, angular momentum, and the energy balance are made, and on the other hand, both the deformation behavior under thermomechanical load and the ability to take and conduct heat under thermal load are described.

Basic treatment of the kinematics, thermomechanical balances, and material theory are well discussed in, e.g., Haupt (2002), Holzapfel (2000), and Bertram (2008). For reasons of consistency, the basic results are listed in this chapter. The mandatory mathematical framework of tensor algebra and analysis used in continuum mechanics is well described in, e.g., Itskov (2013).

3.1 Kinematics

The motion of a body is determined by the time sequence of placements $\chi_t(\mathcal{B})$, where t denotes the progressing time and $\mathcal{B}$ the set of material points $\mathcal{P}$ composing the body. The placement at time t is referred to as current placement. The vector pointing at a material point in the current placement is denoted by $\chi_t(\mathcal{P}) = \boldsymbol{x}(t)$. The initial placement of a body $\boldsymbol{R}(\mathcal{B}) = \chi_{t_0}(\mathcal{B})$, i.e. the placement at the beginning of the process of interest, is called the reference placement. Thereby, the initial time is denoted by $t = t_0$ and usually omitted in these placement related quantities. The vector pointing at the material point in the reference placement is denoted by $\boldsymbol{R}(\mathcal{P}) = \boldsymbol{X}$.

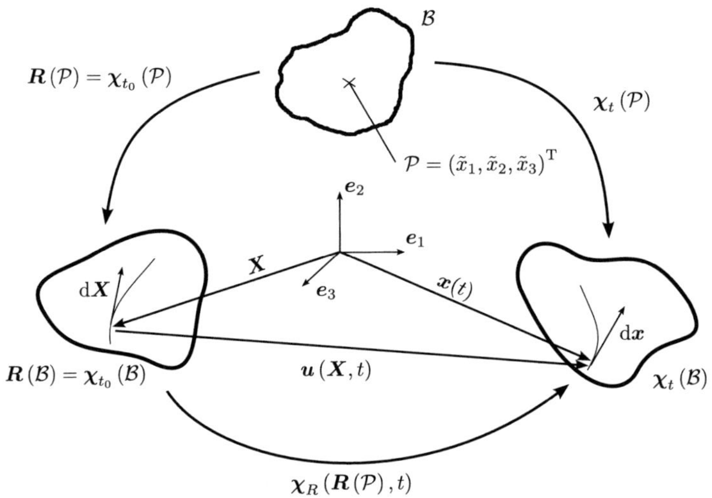

Figure 3.1: Reference placement, current placement, material point, reference based motion

On the basis of the bijective property of the mappings, as depicted in Figure 3.1, all placements can be related to the reference placement by a combination of the mappings $\chi_R(\boldsymbol{R}(\mathcal{B}), t) = \chi_{t_0}^{-1}(\mathcal{B}) \odot \chi_t(\mathcal{B})$. Thus, the position vector of a material point in the current placement can be expressed by

$$\boldsymbol{x}(t) = \chi_R(\boldsymbol{X}, t) = \boldsymbol{u}(\boldsymbol{X}, t) + \boldsymbol{X}, \tag{3.1}$$

where the displacement vector is denoted by $\boldsymbol{u}(\boldsymbol{X}, t)$. Vice versa, the position vector of a material point in the reference placement can be expressed in a similar way

$$\boldsymbol{X} = \boldsymbol{\chi}_R^{-1}(\boldsymbol{x}(t), t) = \boldsymbol{x}(t) - \boldsymbol{u}_e(\boldsymbol{x}(t), t), \tag{3.2}$$

where $\boldsymbol{u}_e(\boldsymbol{x}(t), t) = \boldsymbol{u}(\boldsymbol{\chi}_R^{-1}(\boldsymbol{x}, t), t)$ has been defined. A physical quantity related to the reference placement is called Lagrangian whereas the physical quantity related to the current placement is called Eulerian.

The spatial linearization of the reference placement based motion leads to the definition of the deformation gradient and the displacement gradient

$$\boldsymbol{F}(\boldsymbol{X}, t) = \frac{\partial \boldsymbol{\chi}_R(\boldsymbol{X}, t)}{\partial \boldsymbol{X}} = \operatorname{Grad}(\boldsymbol{\chi}_R(\boldsymbol{X}, t)), \tag{3.3}$$

$$\boldsymbol{H}(\boldsymbol{X}, t) = \frac{\partial \boldsymbol{u}(\boldsymbol{X}, t)}{\partial \boldsymbol{X}} = \operatorname{Grad}(\boldsymbol{u}(\boldsymbol{X}, t)), \tag{3.4}$$

respectively. The connection between these two quantities can be found by the spatial derivation of equation (3.1)

$$\boldsymbol{H}(\boldsymbol{X}, t) = \boldsymbol{F}(\boldsymbol{X}, t) - \boldsymbol{I}. \tag{3.5}$$

Thereby, $\boldsymbol{I}$ is the second-order identity tensor. In the description of the motion of a body, the velocity vector

$$\boldsymbol{v}(\boldsymbol{x}, t) = \dot{\boldsymbol{u}}(\boldsymbol{x}, t) = \frac{\mathrm{D}\boldsymbol{u}(\boldsymbol{x}, t)}{\mathrm{D}t} = \frac{\partial \boldsymbol{u}(\boldsymbol{x}, t)}{\partial t} + \operatorname{grad}(\boldsymbol{u}(\boldsymbol{x}, t))\, \boldsymbol{v} \tag{3.6}$$

and the spatial change of the velocity vector, i.e. the velocity gradient,

$$\boldsymbol{L}(\boldsymbol{x}, t) = \operatorname{grad}(\boldsymbol{v}(\boldsymbol{x}, t)) \tag{3.7}$$

play an important role. Thereby, in the context of a material point in a velocity field, the total time derivation of an arbitrary quantity $\dot{\Xi}$ is referred to as material derivation and marked by $\mathrm{D}\Xi/\,\mathrm{D}t$. The velocity gradient can be decomposed in a uniquely defined symmetric part $\boldsymbol{D}$ called rate of deformation tensor and a skew symmetric part $\boldsymbol{W}$ called spin tensor

$$\boldsymbol{L} = \boldsymbol{D} + \boldsymbol{W}, \tag{3.8}$$

with the definitions

$$\boldsymbol{D} = \frac{1}{2}\left(\boldsymbol{L} + \boldsymbol{L}^{\mathrm{T}}\right) = \boldsymbol{D}^{\mathrm{T}}, \tag{3.9}$$

$$\boldsymbol{W} = \frac{1}{2}\left(\boldsymbol{L} - \boldsymbol{L}^{\mathrm{T}}\right) = -\boldsymbol{W}^{\mathrm{T}}. \tag{3.10}$$

The deformation gradient is multiplicatively decomposable into an orthogonal and a symmetric contribution in two ways

$$\boldsymbol{F} = \boldsymbol{R}\boldsymbol{U} = \boldsymbol{V}\boldsymbol{R}, \tag{3.11}$$

where $\boldsymbol{R}$ denotes the orthogonal orientation tensor and $\boldsymbol{U}$ and $\boldsymbol{V}$ the symmetric right and left Cauchy-Green tensor, respectively. Even though both Cauchy-Green tensors are not affected by rigid body motion, they are not suitable for the description of the strain state of a body. A strain measure is required to be equal zero, if the motion of a body goes without any deformation. Due to this, for $m \in \mathbb{N}$, the tensor valued tensor function

$$\boldsymbol{E}_m^U(\boldsymbol{U}) = \begin{cases} \ln(\boldsymbol{U}) & \text{if} \quad m = 0 \\ m^{-1}\left(\boldsymbol{U}^m - \boldsymbol{I}\right), & \text{if} \quad m \neq 0 \end{cases} \tag{3.12}$$

is introduced as the generalized material strain tensor and the function

$$\boldsymbol{E}_m^V(\boldsymbol{V}) = \begin{cases} \ln(\boldsymbol{V}) & \text{if} \quad m = 0 \\ m^{-1}\left(\boldsymbol{V}^m - \boldsymbol{I}\right), & \text{if} \quad m \neq 0 \end{cases} \tag{3.13}$$

as the generalized spatial strain tensor (Seth, 1964). Some choices of m have special names:

$$\text{Spatial Hencky strain} \quad \boldsymbol{E}_h = \boldsymbol{E}_0^V = \ln(\boldsymbol{V}) = \frac{1}{2}\ln(\boldsymbol{B}) \tag{3.14}$$

$$\text{Material Hencky strain} \quad \boldsymbol{E}_H = \boldsymbol{E}_0^U = \ln(\boldsymbol{U}) = \frac{1}{2}\ln(\boldsymbol{C}) \tag{3.15}$$

$$\text{Green's strain} \quad \boldsymbol{E}_G = \boldsymbol{E}_2^U = \frac{1}{2}(\boldsymbol{U}^2 - \boldsymbol{I}) = \frac{1}{2}(\boldsymbol{C} - \boldsymbol{I}). \tag{3.16}$$

Geometric Linearization

In many technical applications, e.g. the deep drawing step of sheet metal during hot stamping, external loads cause small deformations of a body even if finite rotations of part elements are present. In such a case, the kinematic quantities can be simplified by spatial linearization of the deformation. This leads to a reduction of the infinite number of possible strain measures to one infinitesimal strain quantity.

Given a characteristic length L_0 of the body, the body undergoes small deformations, if and only if

$$\delta = \|\boldsymbol{H}\| \ll 1 \tag{3.17}$$

is valid throughout the entire motion of the body (see, e.g., Haupt, 2002). In the case of small deformations, the kinematic quantities can be linearized with respect to the displacement gradient. This procedure is called geometric linearization.

Remarkable is the fact that, considering small deformations, the infinite number of possible strain measures reduces to one measure

$$\boldsymbol{E}_m^V = \boldsymbol{E}_m^U = \frac{1}{2}\left(\boldsymbol{H} + \boldsymbol{H}^{\mathrm{T}}\right) + \mathbf{0}(\delta^2) = \mathrm{sym}(\mathrm{Grad}\,(\boldsymbol{u})) + \mathbf{0}(\delta^2), \tag{3.18}$$

where $\mathbf{0}(\cdot)$ denotes the Landau symbol. This leads to the definition of the linearized strain tensor or infinitesimal strain tensor

$$\boldsymbol{\varepsilon} = \mathrm{sym}(\mathrm{grad}\,(\boldsymbol{u})) = \frac{1}{2}\left(\boldsymbol{H} + \boldsymbol{H}^{\mathrm{T}}\right). \tag{3.19}$$

3.2 Mechanical and Thermo- dynamical Balance Relations

The deformation of a body is influenced by both the loads from the environment and the forces inside the body referred to as external loads and internal loads, respectively (see, e.g., Haupt, 2002).

Balance relations make statements about the connection between the internal and external loads. In continuum mechanics the mechanical balances, represented by the mass, linear momentum, and angular momentum balance, and the thermomechanical laws, represented by the first and second law of thermodynamics, are considered.

Reynold's Transport Theorem

For moving boundary problems with moving singularities, i.e. for example the transformation front during phase transformation, the Reynold's transport theorem appears to be useful. The Reynold's transport theorem describes the time derivation of an arbitrary additive quantity Ξ in a time-dependent integration volume. In Eulerian form the theorem reads

$$\frac{\mathrm{d}}{\mathrm{d}t}\int\limits_{\chi_t(\mathcal{B})} \Xi\,\mathrm{d}v = \int\limits_{\chi_t(\mathcal{B})} \frac{\partial \Xi}{\partial t}\,\mathrm{d}v + \int\limits_{\partial\chi_t(\mathcal{B})} \Xi\boldsymbol{v}\cdot\boldsymbol{n}\,\mathrm{d}a - \int\limits_{\Omega} [\![\Xi]\!]\boldsymbol{w}\cdot\boldsymbol{n}\,\mathrm{d}a, \tag{3.20}$$

where the singularity, e.g., the transformation front, is denoted by Ω. As depicted in Figure 3.2, the relative velocity of the singularity through

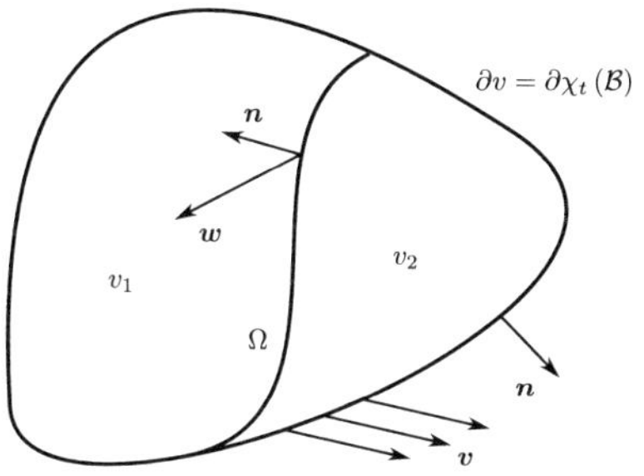

Figure 3.2: Body with moving boundary and singularity

the body is denoted by $\boldsymbol{w}$. $[\![\cdot]\!]$ denotes the difference

$$[\![\Xi]\!] = \Xi_1 - \Xi_2, \tag{3.21}$$

where Ξ_1 is the quantity of the volume in which direction the normal vector of the singularity points.

Balance of Mass

In contrast to the density of a body, which can vary in time for compressible material bodies, the overall mass of a material body is constant throughout the entire motion process

$$\dot{m} = \frac{\mathrm{d}}{\mathrm{d}t} \int\limits_{\chi_t(\mathcal{B})} \rho \, \mathrm{d}v = \frac{\mathrm{d}}{\mathrm{d}t} \int\limits_{\chi_t(\mathcal{R})} \rho_R \, \mathrm{d}V = 0, \tag{3.22}$$

where the density on the reference placement ρ_R is connected with the current density ρ via $\rho = J\rho_R$. Since the local form of the mass balance on the reference placement is trivial, the local form of the mass balance on the current placement reads

$$\dot{\rho} + \rho \mathrm{div}\,(\boldsymbol{v}) = \frac{\partial \rho}{\partial t} + \mathrm{div}\,(\rho \boldsymbol{v}) = 0. \tag{3.23}$$

Balance of Linear Momentum and Angular Momentum

To describe the deformation of a body, the internal and external forces have to be balanced. Newton's second law states that the change in time of the mass' linear momentum is equal to the resulting external force. Considering the body force $\boldsymbol{f}$ and the stress vector $\boldsymbol{t}$, which is related with the Cauchy stress tensor $\boldsymbol{\sigma}$ via Cauchy's theorem $\boldsymbol{t} = \boldsymbol{\sigma n}$, the balance equation in the current placement reads

$$\frac{\mathrm{d}}{\mathrm{d}t} \int\limits_{\chi_t(\mathcal{B})} \boldsymbol{v} \rho \, \mathrm{d}v = \int\limits_{\chi_t(\mathcal{B})} \dot{\boldsymbol{v}} \rho \, \mathrm{d}v = \int\limits_{\partial\chi_t(\mathcal{B})} \boldsymbol{\sigma n} \, \mathrm{d}a + \int\limits_{\chi_t(\mathcal{B})} \boldsymbol{f} \rho \, \mathrm{d}v. \tag{3.24}$$

Assuming a stationary state and using Gauss' theorem, the local form of the linear momentum balance can be derived for the current placement

$$\mathrm{div}\,(\boldsymbol{\sigma}) + \rho \boldsymbol{f} = \boldsymbol{0}. \tag{3.25}$$

The balance of the angular momentum in the current placement yields the important result that the Cauchy stress is symmetric

$$\boldsymbol{\sigma} = \boldsymbol{\sigma}^{\mathrm{T}}. \tag{3.26}$$

Balance of Energy - First Law of Thermodynamics

In thermomechanical processes, the balance of power or energy provides the possibility of determining the temperature of a body. The kinetic energy, due to rigid body motion and the internal energy, which describes the stored energy in the body due to temperature and deformation, are influenced by the external and internal force power, by the heat transfer through the body's surface, and by energy sources and sinks w, e.g., by released or consumed heat during a phase transformation. The heat transfer via the body's surface is given by the normal component of the heat flux vector $\boldsymbol{q}$ at the body's surface. With the balance of linear momentum, and Gauss' theorem, the balance of the internal energy e

reads

$$\frac{\mathrm{d}}{\mathrm{d}t}\int\limits_{\chi_t(\mathcal{B})} e\rho\,\mathrm{d}v = \int\limits_{\chi_t(\mathcal{B})} \boldsymbol{\sigma}\cdot\boldsymbol{D}\,\mathrm{d}v - \int\limits_{\partial\chi_t(\mathcal{B})} \boldsymbol{q}\cdot\boldsymbol{n}\,\mathrm{d}a + \int\limits_{\chi_t(\mathcal{B})} w\rho\,\mathrm{d}v \tag{3.27}$$

Using Gauss' theorem, the local form of the balance equations in the current placement is given by

$$\rho\dot{e} = -\mathrm{div}\,(\boldsymbol{q}) + \boldsymbol{\sigma}\cdot\boldsymbol{D} + \rho w. \tag{3.28}$$

Second Law of Thermodynamics and Clausius-Duhem Inequality

The second law of thermodynamics states that, in a closed system, the change of the entropy is positive. Assuming the entropy supply to be the heat transport by the absolute temperature, what is a good approximation for processes close to equilibrium (see, e.g., Haupt, 2002), the entropy inequality on the current placement

$$\rho\,\dot{\eta} + \mathrm{div}\left(\frac{\boldsymbol{q}}{\theta}\right) - \frac{\rho w}{\theta} \geq 0 \tag{3.29}$$

has to hold for each thermomechanical treatment of a body. Introducing the temperature concerning Legendre-Fenchel transformed quantity of the internal energy, which is referred to as Helmholtz free energy,

$$\psi = e - \theta\eta \tag{3.30}$$

and using the local form of the energy balance equations (3.28), the second law of thermodynamics reads

$$-\rho\dot{\psi} - \rho\eta\dot{\theta} + \boldsymbol{\sigma}\cdot\boldsymbol{D} - \frac{1}{\theta}\boldsymbol{q}\cdot\mathrm{grad}\,(\theta) \geq 0. \tag{3.31}$$

In this form the second law of thermodynamics is referred to as Clausius-Duhem (CD) inequality. A constitutive law fulfilling this inequality is called thermodynamically consistent.

Geometrical Linearization

The geometrical linearization of the balance equations, i.e. the approximation of the balance equations under the condition (3.17), yields

$$\operatorname{div}(\boldsymbol{\sigma}) + \boldsymbol{f} = \mathbf{0} \tag{3.32}$$

$$-\operatorname{div}(\boldsymbol{q}) + \boldsymbol{\sigma} \cdot \dot{\boldsymbol{\varepsilon}} + w = \rho \dot{e} \tag{3.33}$$

$$-\rho\dot{\psi} - \rho\eta\dot{\theta} + \boldsymbol{\sigma} \cdot \dot{\boldsymbol{\varepsilon}} - \frac{1}{\theta}\boldsymbol{q} \cdot \operatorname{grad}(\theta) \geq 0. \tag{3.34}$$

In a small deformation process, a distinction is not made between any placements.

3.3 Time Derivative of Averaged Quantities

In a mean field theory, the modeling is based on the volume averages of the material describing quantities. The volume average of an arbitrary quantity Ξ is defined by

$$\langle \Xi \rangle = \frac{1}{|v|} \int\limits_{v} \Xi \, \mathrm{d}v, \tag{3.35}$$

where the set of the microstructure and the volume of this set are denoted by v and $|v|$, respectively. Volumes of equal thermomechanical properties are referred to as phases. Assuming phase-wise constant values of Ξ, the distribution of the quantity Ξ on the microstructure reads

$$\Xi(\boldsymbol{x}) = \sum_{\alpha \in \mathcal{PH}} \Xi_\alpha \chi_\alpha(\boldsymbol{x}, t) \tag{3.36}$$

with $\mathcal{PH}$ denoting the set of different phases, $\chi_\alpha(\boldsymbol{x})$ the indicator function, and Ξ_α the phase volume average of the quantity Ξ. The indicator function is defined by

$$\chi_\alpha(\boldsymbol{x}, t) = \begin{cases} 1, & \text{if } \boldsymbol{x} \text{ is in phase } \alpha \\ 0, & \text{if otherwise} \end{cases} \tag{3.37}$$

and the phase volume average of a quantity, e.g., Ξ, by

$$\Xi_\alpha = \langle \Xi \rangle_\alpha = \frac{1}{|v_\alpha|} \int_{v_\alpha} \Xi \, \mathrm{d}v, \tag{3.38}$$

where the set of the phase α and the volume of this set are denoted by v_α and $|v_\alpha|$, respectively. Introducing the volume fraction

$$c_\alpha = \frac{|v_\alpha|}{|v|} \tag{3.39}$$

and using equation (3.36), the volume average of a phase-wise constant quantity Ξ reads

$$\langle \Xi \rangle = \sum_{\alpha \in \mathcal{PH}} c_\alpha \Xi_\alpha. \tag{3.40}$$

In the case of the decomposition of a parent phase into N phases as depicted in Figure 3.3 for three existent phases ($N = 2$), the time derivative of a volume averaged quantity reads

$$\frac{\mathrm{d}}{\mathrm{d}t} \langle \Xi \rangle = \langle \dot{\Xi} \rangle + \sum_{\alpha \in \mathcal{PH}} \dot{c}_\alpha \Xi_\alpha. \tag{3.41}$$

Under the assumptions:

- A distinct phase, referred to as parent phase, is transformed into N phases,
- The total volume of the microstructure is constant,
- The boundary between two growing phases does not move,
- Phase-wise constant material behavior, and

- Spatially constant velocity of the transformation front between the parent phase and a growing phase,

relation (3.40) can be found. Using Reynold's Transport Theorem (3.20), the time derivative of $\langle \Xi \rangle$ reads

$$\frac{\mathrm{d}}{\mathrm{d}t} \frac{1}{|v|} \int_{v} \Xi \, \mathrm{d}v = \frac{1}{|v|} \int_{v} \dot{\Xi} \, \mathrm{d}v + \frac{1}{|v|} \int_{\partial v} \Xi \boldsymbol{v} \cdot \boldsymbol{n} \, \mathrm{d}a - \frac{1}{|v|} \int_{\Omega} [\![\Xi]\!] \boldsymbol{w} \cdot \boldsymbol{n} \, \mathrm{d}a. \quad (3.42)$$

Since the microstructure's boundary velocity $\boldsymbol{v}$ is zero and the relative velocity of the singular surface $\boldsymbol{w}_{1\alpha}$ is assumed to be spatially constant on the grain boundary between the parent phase and the growing phase α, the balance equation reads

$$\frac{\mathrm{d}}{\mathrm{d}t} \frac{1}{|v|} \int_{v} \Xi \, \mathrm{d}v = \langle \dot{\Xi} \rangle - \frac{1}{|v|} \sum_{\alpha \in \mathcal{PH} \setminus \{1\}} \int_{\Omega_{1\alpha}} [\![\Xi]\!]_{\alpha} \boldsymbol{w}_{1\alpha} \cdot \boldsymbol{n} \, \mathrm{d}a. \quad (3.43)$$

Hereby, the set of the different phases is $\mathcal{PH} = \{1, \dots, N+1\}$, where, without loss of generality, the distinct parent phase is set to be the first phase. The transformation front of phase α is denoted by $\Omega_{1\alpha}$. Since Ξ_{α} is assumed to be spatially constant on the transformation front $\Omega_{1\alpha}$, the time derivative of an averaged quantity reads

$$\frac{\mathrm{d}}{\mathrm{d}t} \frac{1}{|v|} \int_{v} \Xi \, \mathrm{d}v = \langle \dot{\Xi} \rangle - \sum_{\alpha \in \mathcal{PH} \setminus \{1\}} \left(\Xi_1 - \Xi_{\alpha} \right) \dot{c}_{\alpha}, \quad (3.44)$$

where the relation

$$\dot{c}_{\alpha} = \frac{1}{|v|} \int_{\Omega_{1\alpha}} \boldsymbol{w}_{1\alpha} \cdot \boldsymbol{n} \, \mathrm{d}a = \frac{\dot{|v_{\alpha}|}}{|v|} \quad (3.45)$$

has been used. Introducing the relation

$$\sum_{\alpha \in \mathcal{PH} \setminus 1} \dot{c}_\alpha = -\dot{c}_1 \tag{3.46}$$

into equation (3.44), which can be derived by the time derivative of the sum of the volume fractions $\sum\limits_{\alpha \in \mathcal{PH}} c_\alpha = 1$, equation (3.41) is found.

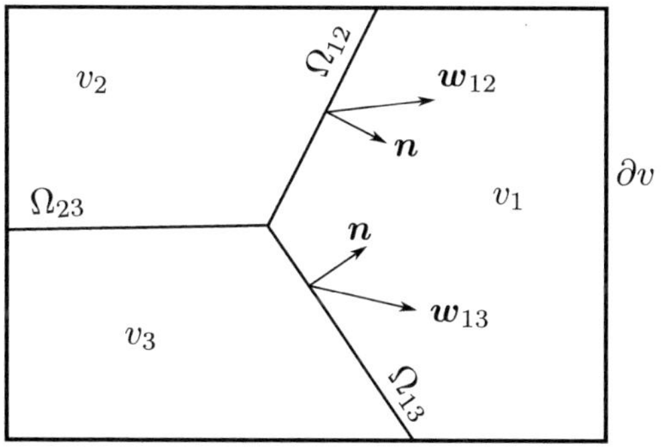

Figure 3.3: Microstructure with a phase transformation of volume 1 into phase 2 and 3

Chapter 4

Constitutive Modeling of the Thermomechanical Behavior[1]

4.1 Potential Relations derived from the Clausius-Duhem Inequality

Assuming an additive decomposition of the infinitesimal strain tensor into an elastic and an inelastic contribution, which represents the plastic, transformation, and or transformation-induced plasticity strain, (see, e.g., Haupt, 2002; Bertram, 2008; Itskov, 2013)

$$\boldsymbol{\varepsilon} = \boldsymbol{\varepsilon}^e + \boldsymbol{\varepsilon}^i, \tag{4.1}$$

and the Helmholtz free energy to be a function of the elastic contribution of the infinitesimal strain tensor $\boldsymbol{\varepsilon}^e$, the temperature, the temperature gradient $\boldsymbol{g} = \mathrm{grad}(\theta)$, and an internal variable, e.g., the accumulated plastic strain ε^p

$$\psi = \hat{\psi}(\boldsymbol{\varepsilon}^e, \theta, \boldsymbol{g}, \varepsilon^p), \tag{4.2}$$

[1] The modeling approach, which is described in this chapter, is based on the paper "Hashine-Shtrikman type mean field model for the two-scale simulation of the thermomechanical processing of steel" (Neumann and Böhlke, 2016) and is extended by a model which takes the TRIP-effect into account.

the Clausius-Duhem inequality (3.34) reads (see, e.g., Coleman and Noll, 1963)

$$\left(\boldsymbol{\sigma} - \rho\frac{\partial\psi}{\partial\boldsymbol{\varepsilon}^e}\right)\cdot\dot{\boldsymbol{\varepsilon}}^e - \rho\left(\frac{\partial\psi}{\partial\theta} + \eta\right)\dot{\theta} - \rho\frac{\partial\psi}{\partial\boldsymbol{g}}\cdot\dot{\boldsymbol{g}} + \boldsymbol{\sigma}\cdot\dot{\boldsymbol{\varepsilon}}^i - \rho\frac{\partial\psi}{\partial\varepsilon^p}\dot{\varepsilon}^p - \frac{1}{\theta}\boldsymbol{q}\cdot\boldsymbol{g} \geq 0. \tag{4.3}$$

For an arbitrary thermomechanical process, this inequality is valid if and only if the potential relations

$$\text{(i) } \boldsymbol{\sigma} = \rho\frac{\partial\psi}{\partial\boldsymbol{\varepsilon}^e}, \qquad \text{(ii) } \eta = -\frac{\partial\psi}{\partial\theta}, \qquad \text{(iii) } \frac{\partial\psi}{\partial\boldsymbol{g}} = \boldsymbol{0} \tag{4.4}$$

hold. The latter relation implies the independence of the Helmholtz free energy of the temperature gradient. The remaining part of equation (4.3) is referred to as the dissipation equation

$$D = D^m + D^\theta \geq 0, \tag{4.5}$$

where the thermal and the mechanical dissipation

$$\text{(i) } D^\theta = -\frac{1}{\theta}\boldsymbol{q}\cdot\boldsymbol{g}, \qquad \text{(ii) } D^m = \boldsymbol{\sigma}\cdot\dot{\boldsymbol{\varepsilon}}^i - \rho\frac{\partial\psi}{\partial\varepsilon^p}\dot{\varepsilon}^p, \tag{4.6}$$

respectively, have been defined. With equation (3.30), the implication of this result on the heat equation (3.33) reads

$$\rho\kappa\dot{\theta} = \boldsymbol{\sigma}\cdot\dot{\boldsymbol{\varepsilon}}^i + \theta\frac{\partial}{\partial\theta}\left(\boldsymbol{\sigma}\cdot\dot{\boldsymbol{\varepsilon}}^e\right) - \left(\iota - \theta\frac{\partial\iota}{\partial\theta}\right)\dot{\varepsilon}^p - \operatorname{div}(\boldsymbol{q}) + w, \tag{4.7}$$

where the abbreviation $\iota = \rho\partial\psi/\partial\varepsilon^p$ and the specific heat capacity

$$\kappa = -\theta\frac{\partial^2\psi}{\partial\theta^2} \tag{4.8}$$

have been introduced. For a wide class of materials, Fourier's law

$$\boldsymbol{q} = -\lambda\boldsymbol{g}. \tag{4.9}$$

is a good assumption for the heat transfer vector. Thereby, $\boldsymbol{\lambda}$ denotes the thermal conductivity tensor. With this constitutive approach, the thermal dissipation $D^\theta \geq 0$ is fulfilled for each thermal process. Introducing the Fourier's law into the heat equation, one obtains

$$\rho\kappa\dot{\theta} - \operatorname{div}\left(\boldsymbol{\lambda}\operatorname{grad}\left(\theta\right)\right) = \boldsymbol{\sigma} \cdot \dot{\boldsymbol{\varepsilon}}^i - \left(\iota - \theta\frac{\partial\iota}{\partial\theta}\right)\dot{\varepsilon}^p + \theta\frac{\partial\boldsymbol{\sigma}}{\partial\theta} \cdot \dot{\boldsymbol{\varepsilon}}^e + w. \quad (4.10)$$

4.2 The Helmholtz Free Energy

The Helmholtz Free Energy in a Small Deformation Framework

For the formulation of the thermomechanical behavior of the material in context of small deformation, the Helmholtz free energy

$$\rho\psi = \frac{1}{2}\left(\boldsymbol{\varepsilon} - \boldsymbol{\varepsilon}^E\right) \cdot \mathbb{C}\left[\boldsymbol{\varepsilon} - \boldsymbol{\varepsilon}^E\right] - \frac{1}{2}\boldsymbol{\alpha} \cdot \mathbb{C}\left[\boldsymbol{\alpha}\right]\Delta\theta^2 + \rho\kappa^c\left(\Delta\theta - \theta\ln\left(\frac{\theta}{\theta_0}\right)\right) + \rho\psi^i\left(\varepsilon^p\right). \quad (4.11)$$

is derived by the introduction of the assumptions (see for the one dimensional case, e.g., Rosakis et al., 2000)

(i) the specific heat capacity is constant,

(ii) the linear elastic relation is unaffected by the inelastic deformation, and

(iii) the thermomechanical stress response is linear in the infinitesimal strain tensor.

Hereby, the fourth-order stiffness tensor is denoted by $\mathbb{C}$, the thermal expansion tensor by $\boldsymbol{\alpha}$, the difference of the current temperature and the reference temperature θ_0 by $\Delta\theta$, and the inelastic energy contribution as a function of the accumulated plastic strain ε^p, also referred to as plastic arc length, by $\psi^i(\varepsilon^p)$. Furthermore, $\boldsymbol{\varepsilon}^E$ denotes the eigenstrain tensor

$$\boldsymbol{\varepsilon}^E = \boldsymbol{\varepsilon}^i + \boldsymbol{\alpha}\Delta\theta = \boldsymbol{\varepsilon}^p + \boldsymbol{\varepsilon}^{tr} + \boldsymbol{\varepsilon}^t + \boldsymbol{\alpha}\Delta\theta, \tag{4.12}$$

which results from an additive decomposition of the inelastic strain into the plastic strain, the TRIP strain $\boldsymbol{\varepsilon}^{tr}$, and the transformation strain $\boldsymbol{\varepsilon}^t$. Note, that the thermal strain is considered as an elastic strain. Also note, that as a result of this approach, the internal energy, which is obtained by a Legendre-Fenchel transformation of the Helmholtz free energy with respect to temperature, as well as the entropy, are additively composed of an elastic, a temperature-dependent, and a hardening contribution. In the following, the statements concerning the Helmholtz free energy, the internal energy, and the entropy made above are derived. With the definition of the specific heat capacity (4.8) and assumption (i), one obtains the following differential equations

$$(a) - \theta\frac{\partial^2\psi}{\partial\theta^2} = \kappa, \qquad (b)\,\frac{\partial^3\psi}{\partial\theta^2\partial\boldsymbol{\varepsilon}^e} = \mathbf{0}, \qquad (c)\,\frac{\partial^3\psi}{\partial\theta^2\partial\varepsilon^p} = \mathbf{0}. \tag{4.13}$$

Assuming enough smoothness of ψ, the integration of (b) and (c) with respect to temperature leads to

$$\rho\frac{\partial\psi}{\partial\boldsymbol{\varepsilon}^e} = \theta\tilde{\boldsymbol{f}}\left(\boldsymbol{\varepsilon}^e, \varepsilon^p\right) + \tilde{\boldsymbol{g}}\left(\boldsymbol{\varepsilon}^e, \varepsilon^p\right), \tag{4.14}$$

$$\rho\frac{\partial\psi}{\partial\varepsilon^p} = \theta\tilde{\tilde{f}}\left(\boldsymbol{\varepsilon}^e, \varepsilon^p\right) + \tilde{\tilde{g}}\left(\boldsymbol{\varepsilon}^e, \varepsilon^p\right). \tag{4.15}$$

Since the linear elastic response is unaffected by the plastic effect (see assumption (ii)), the stress is independent of ε^p

$$\frac{\partial\boldsymbol{\sigma}}{\partial\varepsilon^p} = \rho\frac{\partial^2\psi}{\partial\boldsymbol{\varepsilon}^e\partial\varepsilon^p} = \theta\frac{\partial\tilde{\tilde{f}}}{\partial\boldsymbol{\varepsilon}^e} + \frac{\partial\tilde{\tilde{g}}}{\partial\boldsymbol{\varepsilon}^e} = \mathbf{0}, \tag{4.16}$$

where the result derived from the Clausius-Duhem inequality (4.4)(i) has been used. This equation is valid if and only if both functions, $\tilde{\tilde{f}}$ and

$\tilde{\tilde{g}}$, are independent of the infinitesimal strain tensor. With the potentials $\tilde{\tilde{\eta}}$ and $\tilde{\tilde{e}}$ of $\tilde{\tilde{f}}$ and $\tilde{\tilde{g}}$, respectively,

$$\tilde{\tilde{f}} = -\rho \frac{\partial \tilde{\tilde{\eta}}}{\partial \varepsilon^p}, \qquad \tilde{\tilde{g}} = \rho \frac{\partial \tilde{\tilde{e}}}{\partial \varepsilon^p}, \tag{4.17}$$

the integration of equation (4.15) reads

$$\rho\psi\left(\boldsymbol{\varepsilon}^e, \theta, \varepsilon^p\right) = \tilde{\Psi}\left(\boldsymbol{\varepsilon}^e, \theta\right) - \rho\theta\tilde{\tilde{\eta}}\left(\varepsilon^p\right) + \rho\tilde{\tilde{e}}\left(\varepsilon^p\right). \tag{4.18}$$

Applying the same arguments as before to equation (4.14), one obtains the independence of the functions $\tilde{\boldsymbol{f}}$ and $\tilde{\boldsymbol{g}}$ of the accumulated plastic strain. By introduction of the potentials

$$\tilde{\boldsymbol{f}} = -\rho \frac{\partial \tilde{\eta}}{\partial \boldsymbol{\varepsilon}^e}, \qquad \tilde{\boldsymbol{g}} = \rho \frac{\partial \tilde{e}}{\partial \boldsymbol{\varepsilon}^e} \tag{4.19}$$

and by integration of the differential equation

$$\frac{\partial \tilde{\Psi}}{\partial \boldsymbol{\varepsilon}^e} = \theta \tilde{\boldsymbol{f}}\left(\boldsymbol{\varepsilon}^e\right) + \tilde{\boldsymbol{g}}\left(\boldsymbol{\varepsilon}^e\right) \tag{4.20}$$

gained by use of equation (4.18) and (4.14), one obtains

$$\tilde{\Psi} = \Psi\left(\theta\right) - \rho\theta\tilde{\eta}\left(\boldsymbol{\varepsilon}^e\right) + \rho\tilde{e}\left(\boldsymbol{\varepsilon}^e\right) \tag{4.21}$$

and the Helmholtz free energy

$$\rho\psi = \Psi\left(\theta\right) - \rho\theta\left(\tilde{\eta}\left(\boldsymbol{\varepsilon}^e\right) + \tilde{\tilde{\eta}}\left(\varepsilon^p\right)\right) + \rho\left(\tilde{e}\left(\boldsymbol{\varepsilon}^e\right) + \tilde{\tilde{e}}\left(\varepsilon^p\right)\right). \tag{4.22}$$

The potential relation of the entropy (4.4).(ii) allows a physical interpretation of the potentials $\tilde{\eta}$ and $\tilde{\tilde{\eta}}$

$$\eta = -\frac{\partial \psi}{\partial \theta} = -\frac{1}{\rho}\frac{\partial \Psi}{\partial \theta} + \tilde{\eta}\left(\boldsymbol{\varepsilon}_G^e\right) + \tilde{\tilde{\eta}}\left(\varepsilon^p\right). \tag{4.23}$$

$\tilde{\eta}\left(\varepsilon_G^e\right)$ arises as the contribution to the entropy due to elasticity and $\tilde{\tilde{\eta}}\left(\varepsilon^p\right)$ as the contribution due to plasticity. Note, that the temperature, the elasticity, and the plasticity contribute additively to the entropy. The same result is obtained for the internal energy. The Legendre-Fenchel transformation of equation (4.22) with respect to temperature reads

$$\rho e = \rho\psi + \theta\eta = \Psi\left(\theta\right) - \theta\frac{\partial\Psi}{\partial\theta} + \rho\tilde{e}\left(\boldsymbol{\varepsilon}^e\right) + \rho\tilde{\tilde{e}}\left(\varepsilon^p\right), \tag{4.24}$$

where the additive nature of the contributions due to temperature, the stored energy due to elasticity, and the stored energy due to plasticity is apparent. To determine the entropy contribution and the stored energy due to elasticity, the restriction on the constitutive law, i.e., assumption (iii), is used, which is equivalent to

$$\rho\frac{\partial^2\psi}{\partial\,\boldsymbol{\varepsilon}^{e\,2}} = -\rho\theta\frac{\partial^2\tilde{\eta}}{\partial\,\boldsymbol{\varepsilon}^{e\,2}} + \rho\frac{\partial^2\tilde{e}}{\partial\,\boldsymbol{\varepsilon}^{e\,2}} = \mathbb{C}. \tag{4.25}$$

This equation is valid for all temperatures if and only if the following conditions

$$\frac{\partial^2\tilde{\eta}}{\partial\,\boldsymbol{\varepsilon}^{e\,2}} = 0, \qquad \rho\frac{\partial^2\tilde{e}}{\partial\,\boldsymbol{\varepsilon}^{e\,2}} = \mathbb{C} \tag{4.26}$$

hold. Setting the stored energy due to elasticity to zero for vanishing elastic strain and assuming linear stress strain response, the integration of both equations read

$$\rho\tilde{\eta} = \boldsymbol{\varepsilon}^e\cdot\mathbb{C}\left[\boldsymbol{\alpha}\right], \qquad \rho\tilde{e} = \frac{1}{2}\boldsymbol{\varepsilon}^e\cdot\mathbb{C}\left[\boldsymbol{\varepsilon}^e\right], \tag{4.27}$$

where the integration constant of the entropy is denoted by $\mathbb{C}\left[\boldsymbol{\alpha}\right]/\rho$ and the thermal expansion coefficient by $\boldsymbol{\alpha}$. The integration of equation (4.13).(a) under consideration of equation (4.22) reads

$$\Psi\left(\theta\right) = -\rho\kappa^c\theta\ln\left(\frac{\theta}{\theta_0}\right) + \rho\kappa^c\Delta\theta, \tag{4.28}$$

if the stored energy due to temperature is zero for a temperature equal to the reference temperature. Under the assumptions (i)-(iii), the resulting Helmholtz free energy and the entropy have to be in the form

$$\rho\psi = \frac{1}{2}\boldsymbol{\varepsilon}^e \cdot \mathbb{C}\left[\boldsymbol{\varepsilon}^e\right] - \boldsymbol{\varepsilon}^e \cdot \mathbb{C}\left[\boldsymbol{\alpha}\right]\Delta\theta + \rho\kappa^c\left(\Delta\theta - \theta\ln\left(\frac{\theta}{\theta_0}\right)\right) + \rho\psi^i\left(\varepsilon^p\right), \tag{4.29}$$

$$\rho\eta = \boldsymbol{\varepsilon}^e \cdot \mathbb{C}\left[\boldsymbol{\alpha}\right] + \rho\kappa^c \ln\left(\frac{\theta}{\theta_0}\right), \tag{4.30}$$

respectively. Introducing the eigenstrain tensor (4.12) into the Helmholtz free energy (4.29), the final form of the Helmholtz free energy in equation (4.11) is obtained. Hereby, the entropy due to plasticity is neglected and the stored energy due to plasticity is denoted by ψ^i.

The resulting stress strain relation finally reads

$$\boldsymbol{\sigma} = \rho\frac{\partial\psi}{\partial\boldsymbol{\varepsilon}^e} = \mathbb{C}\left[\boldsymbol{\varepsilon}^e - \boldsymbol{\alpha}\Delta\theta\right] = \mathbb{C}\left[\boldsymbol{\varepsilon} - \boldsymbol{\varepsilon}^i - \boldsymbol{\alpha}\Delta\theta\right] = \mathbb{C}\left[\boldsymbol{\varepsilon} - \boldsymbol{\varepsilon}^E\right], \tag{4.31}$$

which is referred to as Hooke's law.

Implications on the Heat Equation

To describe the heating of a body due to inelastic effects, the Taylor-Quinney factor (see, e.g., Farren and Taylor, 1925; Taylor and Quinney, 1934) is often used. The Taylor-Quinney factor is the ratio of the rate of the dissipated inelastic work

$$\dot{Q} = \boldsymbol{\sigma} \cdot \dot{\boldsymbol{\varepsilon}}^i - \left(\iota - \rho\theta\frac{\partial\iota}{\partial\theta}\right)\dot{\varepsilon}^p \tag{4.32}$$

and the rate of the total inelastic work $\dot{W} = \boldsymbol{\sigma} \cdot \dot{\boldsymbol{\varepsilon}}^i$

$$\beta = \frac{\dot{Q}}{\dot{W}} = 1 - \frac{\left(\iota - \theta\frac{\partial\iota}{\partial\theta}\right)\dot{\varepsilon}^p}{\boldsymbol{\sigma} \cdot \dot{\boldsymbol{\varepsilon}}^i} \tag{4.33}$$

With equation (4.29), the Taylor-Quinney factor can be reduced to lower terms

$$\beta = 1 - \frac{\rho\dot{\psi}^i}{\boldsymbol{\sigma} \cdot \dot{\boldsymbol{\varepsilon}}^i}. \tag{4.34}$$

This means that the Taylor-Quinney factor indicates how much of the inelastic work is dissipated. For an ideal plastic material, i.e., a material without any plastic hardening, the factor is one and the entire inelastic work is dissipated. For a material with $\beta = 0$, the entire inelastic work is used for the hardening. Often, the Taylor-Quinney factor is used as a material parameter (see, e.g., Kapoor and Nemat-Nasser, 1998; Brünig and Driemeier, 2007; Khan and Liu, 2012), but the mechanical stress/strain behavior determines its value (see, e.g., Rosakis et al., 2000). In terms of the inelastic work and the Taylor-Quinney factor, the heat equation reads

$$\rho\kappa\dot{\theta} - \operatorname{div}\left(\boldsymbol{\lambda}\operatorname{grad}\left(\theta\right)\right) = \beta\dot{W} + Q^e + w, \tag{4.35}$$

with the abbreviation $Q^e = -\theta\mathbb{C}\left[\boldsymbol{\alpha}\right] \cdot \dot{\boldsymbol{\varepsilon}}^e$.

Extension to Temperature-Dependent Elasticity and Hardening

If high temperature changes occur in time, the elastic behavior, as well as the yield surface are observed to be temperature-dependent. Besides the extension of the stiffness tensor $\mathbb{C} = \mathbb{C}(\theta)$, it is assumed that the stored energy of cold work is a function of temperature $\psi^i = \psi^i(\gamma, \theta)$. This impacts the entropy

$$\eta = \frac{1}{\rho}\boldsymbol{\varepsilon}^e \cdot \mathbb{C}\left[\boldsymbol{\alpha}\right] + \kappa^c \ln\left(\frac{\theta}{\theta_0}\right) - \frac{1}{2\rho}\boldsymbol{\varepsilon}^e \cdot \frac{\partial\mathbb{C}}{\partial\theta}\left[\boldsymbol{\varepsilon}^e\right] + \frac{1}{\rho}\boldsymbol{\varepsilon}^e \cdot \frac{\partial\mathbb{C}}{\partial\theta}\left[\boldsymbol{\alpha}\right]\Delta\theta - \frac{\partial\psi^i}{\partial\theta} \tag{4.36}$$

and the specific heat capacity

$$\kappa = \theta\frac{\partial\eta}{\partial\theta} = \kappa^c + \frac{2\theta}{\rho}\boldsymbol{\varepsilon}^e \cdot \frac{\partial\mathbb{C}}{\partial\theta}\left[\boldsymbol{\alpha}\right] - \theta\frac{\partial^2\psi^i}{\partial\theta^2}, \tag{4.37}$$

where κ^c is the constant part of the specific heat capacity. Furthermore, the temperature dependence of the stiffness tensor is assumed to be linear. The validity of the suggested model despite the extension is shown in Chapter 7 by consideration of the second and the third term of the latter equation. It is shown that these terms can be neglected for the both steel grades 42CrMo4 and 22MnB5.

4.3 Plasticity, TRIP, and Transformation Strain

Since only the stored energy of cold work $\psi^i(\varepsilon^p)$ in the Helmholtz free energy (4.11) is dependent on the accumulated plastic strain, the plastic thermodynamic conjugate reads $\iota^p = \rho\partial\psi^i/\partial\varepsilon^p$. With equation (4.12) the plastic part of the mechanical dissipation (4.6)(ii) is defined by

$$D^{m,p} = \boldsymbol{\sigma} \cdot \dot{\boldsymbol{\varepsilon}}^p - \iota^p \dot{\varepsilon}^p \geq 0. \tag{4.38}$$

This is a stricter condition than the original one given in condition (4.6).(ii). The postulate of maximum plastic dissipation and the introduction of a J_2-flow theory corresponding yield surface (see, e.g., Simo and Hughes, 1998; Xiao et al., 2012),

$$\phi(\boldsymbol{\sigma}, \varepsilon^p) = \|\boldsymbol{\sigma}'\| - \sqrt{\frac{2}{3}}\left(\sigma^0 + \iota^p\right) \tag{4.39}$$

leads to the classical Kuhn-Tucker optimality conditions

$$\frac{\partial \mathcal{L}^p}{\partial \boldsymbol{\sigma}} = \mathbf{0}, \quad \frac{\partial \mathcal{L}^p}{\partial \iota^p} = 0 \tag{4.40}$$

with the Lagrangian functional

$$\mathcal{L}^p = -\tilde{D}^{m,p} + \dot{\gamma}\phi(\boldsymbol{\sigma}, \iota^p). \tag{4.41}$$

The Lagrangian functional has to be minimized under the loading/unloading conditions in the Kuhn-Tucker complementary form to get the maximum plastic dissipation

$$\dot{\gamma} \geq 0, \quad \phi\left(\boldsymbol{\sigma}, \varepsilon^p\right) \leq 0, \quad \dot{\gamma}\phi\left(\boldsymbol{\sigma}, \iota^p\right) = 0. \tag{4.42}$$

The minimization of the Lagrangian delivers both the flow rule, i.e. the evolution equation for the plastic strain, and the evolution equation for the accumulated plastic strain

$$\text{(i) } \dot{\boldsymbol{\varepsilon}}^p = \dot{\gamma}\frac{\boldsymbol{\sigma}'}{\|\boldsymbol{\sigma}'\|}, \qquad \text{(ii) } \dot{\varepsilon}^p = \dot{\gamma}\sqrt{\frac{2}{3}}, \tag{4.43}$$

where $\dot{\gamma}$ denotes the consistency parameter. The deviatoric part of the Cauchy stress tensor is denoted by $\boldsymbol{\sigma}'$ and the initial yield stress by σ^0. For a Voce-type hardening, the inelastic stored energy reads

$$\rho\psi^i\left(\varepsilon^p\right) = \frac{1}{2}\Theta^\infty\, {\varepsilon^p}^2 + \left(\sigma^\infty - \sigma^0\right)\left(\varepsilon^p - \frac{\sigma^\infty - \sigma^0}{\Theta^\infty - \Theta^0}\exp\left(\frac{\Theta^\infty - \Theta^0}{\sigma^\infty - \sigma^0}\varepsilon^p\right)\right). \tag{4.44}$$

This approach leads to the yield surface (Buchheit et al., 2005; Jöchen and Böhlke, 2012; Voce, 1955)

$$\phi = \|\boldsymbol{\sigma}'\| - \sqrt{\frac{2}{3}}\left(\sigma^0 + \Theta^\infty\varepsilon^p + \left(\sigma^\infty - \sigma^0\right)\left(1 - \exp\left(\frac{\Theta^\infty - \Theta^0}{\sigma^\infty - \sigma^0}\varepsilon^p\right)\right)\right) \leq 0. \tag{4.45}$$

As an alternative hardening approach, a temperature-dependent Swift-type (Swift, 1952) isotropic hardening is suggested by Bok et al. (2014)

$$\phi = \|\boldsymbol{\sigma}'\| - \sqrt{\frac{2}{3}}s^f(\theta)\left(\varepsilon^0 + \varepsilon^p\right)^{s^e} \leq 0. \tag{4.46}$$

The associated inelastic stored energy reads

$$\rho\psi^i\left(\varepsilon^p\right) = \frac{s^f(\theta)}{s^e(\theta)+1}\left(\varepsilon^0+\varepsilon^p\right)^{s^e+1} - s^f(\theta)\left(\varepsilon^0\right)^{s^e}\varepsilon^p, \tag{4.47}$$

where the temperature-dependent Swift parameters are given by

$$s^f(\theta) = s^{f,0} + s^{f,1}\theta, \qquad s^e(\theta) = s^{e,0} + s^{e,1}\theta. \tag{4.48}$$

If both, the yield condition (4.39) and the loading condition (Bertram and Krawietz, 2012)

$$\frac{\partial\phi}{\partial\boldsymbol{\varepsilon}^e}\cdot\dot{\boldsymbol{\varepsilon}}^e + \frac{\partial\phi}{\partial\theta}\dot{\theta} > 0 \tag{4.49}$$

are fulfilled, the material behavior is elasto-plastic. Note that for a temperature-independent yield surface, the second part of the latter equation vanishes.

The TRIP effect is a macroscopic effect based on the micro-plastic strains in the weaker phase caused by the volume change of the growing phase during phase transformation. Therefore, at this point, the macroscopic Leblond-based modeling of the TRIP effect is considered. The modeling in the context of the two-scale approach is discussed in Chapter 5. Based on a spherical microstructure and the Sachs approximation (see Appendix B.2), Leblond et al. (1989) derived an evolution equation for the TRIP strain

$$\dot{\boldsymbol{\varepsilon}}^{tr} = \frac{3}{2}k^{tr}\varphi'(c_\alpha)\dot{c}_\alpha\boldsymbol{\sigma}'\mathcal{H}(\dot{c}_\alpha), \tag{4.50}$$

where k^{tr} denotes the TRIP parameter, φ the TRIP saturation function, and $\mathcal{H}$ the Heaviside function. With the conditions on the saturation function $\vartheta_\alpha(0) = 0$ and $\vartheta_\alpha(1) = 1$, in literature, different approaches can be found:

$$\text{Wolff et al. (2008a):} \qquad \varphi(c_\alpha) = c_\alpha, \tag{4.51}$$

$$\text{Desalos (1981):} \qquad \varphi(c_\alpha) = c_\alpha\,(2 - c_\alpha)\,, \tag{4.52}$$

$$\text{Leblond et al. (1989):} \qquad \varphi(c_\alpha) = c_\alpha\,(1 - \ln c_\alpha)\,. \tag{4.53}$$

Thereby, the Leblond's saturation function results naturally from a microscopical consideration of an isotropic growth of the new phase in a spherical parent phase. The first saturation function is usually referred to as the Tanaka model. Böck and Holzapfel (2002) showed that Leblond's equation can be regarded as a viscoplastic evolution law of the Perzyna-type (Perzyna, 1971)

$$\dot{\boldsymbol{\varepsilon}}^{tr} = \dot{\gamma}^{tr} \frac{\partial \phi^{tr}}{\partial \boldsymbol{\sigma}}, \tag{4.54}$$

where the consistency parameter is given by

$$\dot{\gamma}^{tr} = \begin{cases} \frac{1}{\epsilon} \frac{\phi^{tr}}{k^{tr}} & \text{if} \quad \phi \geq 0 \\ 0 & \text{if} \quad \phi < 0 \end{cases}. \tag{4.55}$$

The yield surface ϕ^{tr} is a von Mises yield function with zero yield stress $\phi^{tr} = \|\boldsymbol{\sigma}'\|$ and the relaxation time

$$\epsilon = \frac{2}{3\varphi'(c_\alpha)\dot{c}_\alpha \mathcal{H}(\dot{c}_\alpha)} \tag{4.56}$$

goes towards infinity if no phase transformation is present. With $\partial\phi/\partial\boldsymbol{\sigma} = \boldsymbol{\sigma}'/\|\boldsymbol{\sigma}'\| = \boldsymbol{n}$, equation (4.50) is obtained. The implication of the interpretation of Leblond's evolution law as an ideal viscoplastic material behavior without hardening on the dissipation due to TRIP is that the entire mechanical work $\boldsymbol{\sigma} \cdot \dot{\boldsymbol{\varepsilon}}^{tr}$ is dissipated. However, since the TRIP strain is deviatoric, only deviatoric stress contributes to the mechanical work of the TRIP deformation.

During phase transformation, volume changes occur, which are referred to as transformation strains. The transformation strain is determined by the anisotropic volume change of the different variants of the new phase. Assuming a uniform distribution of the variants (see, e.g., Yu et al., 2014b; Wang et al., 2008), the transformation strain can be assumed to be isotropic

$$\varepsilon^t = \zeta \boldsymbol{I} \mathcal{H}\left(t - t_0\right), \tag{4.57}$$

with the mean volume change due to transformation ζ and the transformation start time t_0. Other isotropic approaches, e.g. based on the density, and anisotropic approaches, based on a consideration of the different variants of the growth phase are suggested by Wolff et al. (2008b), and Mahnken et al. (2015), respectively. Similarly to the TRIP effect, the entire mechanical work due to transformation strain is dissipated. However, since the transformation strain is isotropic, only the spherical stress part contributes to the dissipation.

Using equations (4.39), (4.43), (4.50), and (4.57), the mechanical dissipation due to classic plasticity, TRIP, and transformation strain read

$$D^{m,p} = \sigma^0 \dot{\varepsilon}^p, \tag{4.58}$$

$$D^{m,tr} = \frac{3}{2} k^{tr} \varphi'(c_\alpha) \|\boldsymbol{\sigma}'\|^2 \max(\dot{c}, 0), \tag{4.59}$$

$$D^{m,t} = \zeta \mathrm{tr}\left(\boldsymbol{\sigma}\right) \delta\left(t_0\right) \tag{4.60}$$

leading to a positive total mechanical dissipation

$$D^m = \sigma^0 \dot{\varepsilon}^p + \frac{3}{2} k^{tr} \varphi'(c_\alpha) \|\boldsymbol{\sigma}'\|^2 \max(\dot{c}, 0) + \zeta \mathrm{tr}\left(\boldsymbol{\sigma}\right) \delta\left(t_0\right) \geq 0. \tag{4.61}$$

Thereby, $\delta(\cdot)$ denotes the Dirac delta distribution and tr $(\cdot)$ the trace of a second-order tensor. For an isotropic linear elastic behavior

$$\mathbb{C} = 3K\mathbb{P}^{1,\mathrm{iso}} + 2G\mathbb{P}^{2,\,\mathrm{iso}}, \tag{4.62}$$

the heat equation (4.10) can be given in the form

$$\rho\kappa\dot{\theta} - \operatorname{div}\left(\boldsymbol{\Lambda}\mathrm{grad}\,(\theta)\right) = \sigma^0\dot{\varepsilon}^p + \frac{3}{2}k^{tr}\varphi'(c_\alpha)\|\boldsymbol{\sigma}'\|^2\max(\dot{c},0) + \zeta\mathrm{tr}\,(\boldsymbol{\sigma})\delta\,(t_0) + 3\rho\alpha K\theta\mathrm{tr}\,(\dot{\boldsymbol{\varepsilon}}^e) + w. \tag{4.63}$$

Thereby, K denotes the bulk modulus, G the shear modulus, $\mathbb{P}^{1,\,\mathrm{iso}}$ the first fourth-order projector, and $\mathbb{P}^{2,\,\mathrm{iso}}$ the second fourth-order projector defined by

$$\mathbb{P}^{1,\,\mathrm{iso}} = \frac{1}{3}\boldsymbol{I}\otimes\boldsymbol{I}, \qquad \mathbb{P}^{2,\,\mathrm{iso}} = \mathbb{I}^s - \mathbb{P}^{1,\,\mathrm{iso}}, \tag{4.64}$$

where $\mathbb{I}^s$ is the fourth-order identity tensor on symmetric second-order tensors.

4.4 Constitutive Laws of Phase Transformation

4.4.1 The Diffusion-Driven Decomposition of Austenite

In the cooling process of an austenitized sheet metal, ferrite, pearlite, and bainite arise as a result of the diffusion of alloy elements in the austenitic parent phase. During the thermal treatment, the austenitic FCC lattice flips into a carbon saturated BCC lattice referred to as ferrite. Carbon is deposited due to diffusion form a pearlitic or bainitic structure. The phases pearlite and bainite differ in shape, size and distribution. Due to intermediate cooling rates, the diffusion of carbon is not completed and metastable bainitic microstructures occur. In literature, the bainitic phase transformation is often treated like a purely diffusion driven phase transformation (see, e.g., Miokovic, 2005; Wolff et al., 2008b; De Oliveira et al., 2010). Furthermore, there are also different approaches based on thermodynamical arguments as suggested by, e.g., Mahnken et al. (2015),

In literature, there are two methods used to model the diffusion-driven phase transformation: the Johnson-Mehl-Avrami-Kolmogorov (JMAK) and the Kirkaldy-Venugopalan (KV) approach.

Isothermal Phase Transformation Model of JMAK

The JMAK approach is based on an extended volume approach (see, e.g., Kolmogorov, 1937; Avrami, 1939). Assuming an infinitely large parent phase, the nucleated germs form the extended volume taking the overlapping volumes into account occurring during growth of the supercritical germs. For a given sample volume V which is supposed to be constant throughout the transformation, a nucleation rate $\dot{I}_\alpha(t)$, and a growth velocity $Y_\alpha(t)$ of a phase α, the extended volume fraction of that phase is determinable by

$$c_\alpha^e = \frac{|V_\alpha^e|}{|V|} = \int_0^t \left[g_\alpha \int_\tau^t Y_\alpha(\varsigma)\, \mathrm{d}\varsigma \right]^m \dot{I}_\alpha(\tau)\, \mathrm{d}\tau. \tag{4.65}$$

The internal integral of equation (4.65) represents the volume of a supercritical germ at time t nucleated at τ with m being the dimension parameter and g_α the geometry factor. If the distribution of the subcritical embryos is random, the volume fraction of a phase α is able to be derived by the consideration of the volume change of the new phase $\mathrm{d}V_\alpha$. During phase transformation, only a part of the extended volume change $\mathrm{d}V_\alpha^e$ contributes to the volume change of the new phase. This part is the untransformed volume fraction (see, e.g. Kolmogorov, 1937). In multi-component steels the untransformed volume fraction c_α^u is determined by

$$c_\alpha^u = c_\alpha^r - c_\alpha, \tag{4.66}$$

where c^r_α denotes the remaining volume fraction of austenite before the transformation of phase α starts

$$c^r_\alpha = 1 - \sum_{\substack{\beta=1 \\ \beta\neq\alpha}}^{N} c_\beta . \tag{4.67}$$

The volume change of the new phase is given by

$$|\,\mathrm{d}V_\alpha| = c^u_\alpha |\,\mathrm{d}V^e_\alpha|, \tag{4.68}$$

leading to the differential equation for the determination of the volume fraction of phase α after some rearrangements

$$c'_\alpha = \frac{\mathrm{d}c_\alpha}{\mathrm{d}c^e_\alpha} = c^r_\alpha \left(1 - c_\alpha\right) \tag{4.69}$$

with the solution

$$c_\alpha = c^r_\alpha \left(1 - \exp\left(-c^e_\alpha\right)\right). \tag{4.70}$$

In the case of anisotropic growth of the phase in the parent phase, hard impingement due to mutual blocking effect of the growing supercritical germs arises. Due to this, the part from the extended volume change contributing to the volume change of the new phase has to be modified. A standard approach is given by $c'_\alpha = {c^u_\alpha}^\xi$ with the solution c_α (see, e.g., Starink and Zahra, 1998; Liu et al., 2007)

$$c_\alpha = c^r_\alpha \left[{c^r_\alpha}^{1-\xi} - (1-\xi)\, c^e_\alpha \right]^{\frac{1}{1-\xi}}, \tag{4.71}$$

where $1/(1-\xi)$ is referred to as the impingement exponent. With this exponent, the kind of impingement can be influenced.
For the description of the nucleation, three basic models are used: saturation by pre-existing supercritical germs, continuous nucleation, and Avrami nucleation. In the case of saturation, the nucleation rate of phase

α is given by $\dot{I}_\alpha(\theta) = I_\alpha^{sat}\delta(t_0)$, where I_α^{sat} denotes the saturation. At the start of the transformation, the number of supercritical germs I_α is equal to I_α^{sat} and stays constant throughout the process.

In the continuous case, for large undercooling or overheating, the temperature dependence of the nucleation rate is assumed to obey an Arrhenius law

$$\dot{I}_\alpha(\theta) = I_\alpha^{con} \exp\left(-\frac{Q_\alpha^I}{R\theta}\right) \tag{4.72}$$

with Q_α^I being the activation energy of nucleation, R the universal gas constant, and I_α^{con} the nucleation rate of phase α at temperature going against zero (Liu et al., 2007). Assuming an isothermal process, equation (4.72) is analytically integrable and the number of supercritical germs I_α increases linearly with time.

The Avrami nucleation is based on constant sum of subcritical embryos I_α^{sub} and supercritical germs. Furthermore, it is assumed that the change of the subcritical embryos is proportional to the number of the subcritical embryos. Altogether, the nucleation rate of phase α is given by $\dot{I}_\alpha = -\dot{I}_\alpha^{sub} = r_\alpha I_\alpha^{sub}$ with r_α denoting the transition rate into nucleated germs. With an Arrhenius ansatz for the transition rate into nucleated germs, the solution of this differential equation provides an equation for the nucleation rate

$$\dot{I}_\alpha(\theta) = r_\alpha I_\alpha^{sat} \exp\left(\int_0^t r_\alpha^0 \exp\left(-\frac{Q_\alpha^I}{R\theta}\right) \mathrm{d}\tau\right), \tag{4.73}$$

with r_α^0 being a temperature-independent reference transition rate.

The different nucleation rates are depicted in Figure 4.1, where $\theta = 600\mathrm{K}$, $I_\alpha^{sat} = 2 \cdot 10^{21} 1/\mathrm{m}^3$, $I_\alpha^{con} = 10^{40} 1/\mathrm{m}^3\mathrm{s}$, $Q_\alpha^I = 257000\mathrm{J/mol}$, and the three different transition rates $\iota_{\alpha,1}^o = 8 \cdot 10^{18} 1/\mathrm{m}^3 s$, $\iota_{\alpha,2}^o = 6 \cdot 10^{19} 1/\mathrm{m}^3 s$, and $\iota_{\alpha,1}^o = 3 \cdot 10^{20} 1/\mathrm{m}^3 s$ have been used.

For large undercooling or overheating, the growth velocity of phase α

can be expressed by

$$Y_\alpha(\theta) = Y_\alpha^0 \exp\left(-\frac{Q_\alpha^Y}{R\theta}\right), \tag{4.74}$$

where Y_α^0 denotes the growth rate at a temperature going against zero and Q_α^Y the activation energy of growth (Liu et al., 2007).

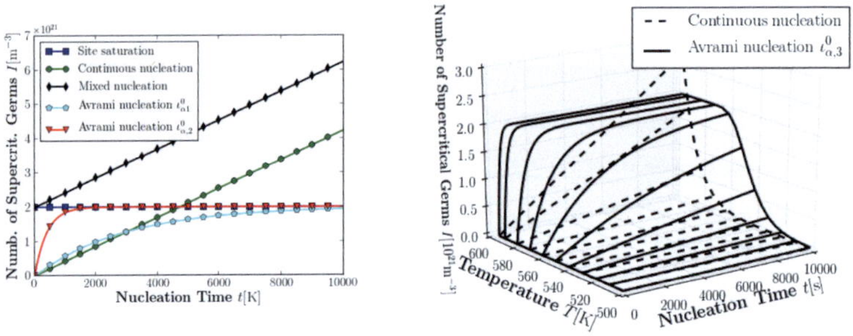

Figure 4.1: Comparison of different kinds of nucleation

However, the JMAK equation can be derived by assuming that (see, e.g., Starink, 2001)

(i) The initial state of the sample is homogeneous,
(ii) The phase transformation takes place under isothermal conditions,
(iii) Subcritical embryos are randomly distributed,
(iv) The nucleation is continuous,
(v) Growth is isotropic and the growth rate is independent of position in the sample, and that
(vi) Hard impingement due to blocking effect is negligible.

With these assumptions, using equation (4.70), (4.65), (4.72), and (4.74), the JMAK equation reads (see, e.g., Krüger, 1993)

$$c_\alpha = c_\alpha^r \left(1 - \exp\left(-b_\alpha(\theta)\, t^{a_\alpha(\theta)}\right)\right), \tag{4.75}$$

where the JMAK parameters

$$b_\alpha(\theta) = \frac{g}{m+1} I_\alpha^{con}\, {Y_\alpha^0}^m \exp\left(-\frac{mQ_\alpha^Y + Q_\alpha^I}{R\theta}\right) \tag{4.76}$$

and $a_\alpha = m + 1$ have been introduced. Following the approach of Nußkern (2013), a more flexible model for the JMAK parameter b_α is suggested

$$b_\alpha = b_{1,\alpha} \exp\left(-\frac{(\theta - b_{2,\alpha})^2}{2b_{3,\alpha}^2}\right). \tag{4.77}$$

Furthermore, for a better agreement with experimental results, polynomials of up to fifth order are often used for the Avrami parameters

$$b_\alpha(\theta) = a_\alpha^b + b_\alpha^b\theta + c_\alpha^b\theta^2 + d_\alpha^b\theta^3 + e_\alpha^b\theta^4 + f_\alpha^b\theta^5 + g_\alpha^b\theta^6 \tag{4.78}$$

$$a_\alpha(\theta) = a_\alpha^a + b_\alpha^a\theta + c_\alpha^a\theta^2 + d_\alpha^a\theta^3 + e_\alpha^a\theta^4 + f_\alpha^a\theta^5 + g_\alpha^a\theta^6 \tag{4.79}$$

as proposed in, e.g., Hömberg (1996), Miokovic (2005), and De Oliveira et al. (2010). Due to a higher amount of parameters in this approach, the result has a greater flexibility of the transformation behavior at the expense of physical interpretability of the transformation parameter.

In the heating case, i.e. the austenitization of an initial ferritic microstructure, the JMAK parameters are chosen as

$$b_a(\theta) = a_a^b \exp\left(-\frac{b_a^b}{c_a^b\theta}\right), \quad a_a(\theta) = a_a^a + b_a^a\theta. \tag{4.80}$$

Isothermal Phase Transformation Model of KV

The Kirkaldy-Venugopalan (KV) model is often used to describe the diffusion-driven polymorphic steel behavior in the quenching step of the hot stamping process (see, e.g., Ackerström et al., 2007; Olsson, 2009; Hochholdinger et al., 2011). Kirkaldy and Venugopalan (1983) proposed a model for the evolution of the volume fractions in an isothermal process, which is based on a multiplicative decomposition of the quantities influencing the phase transformation

$$\dot{\tilde{c}}_\alpha = \dot{\tilde{c}}_\alpha(\tilde{c}_\alpha, \theta, S, \mathcal{C}) = f_\alpha^S(S) f_\alpha^\theta(\theta) f_\alpha^{\mathcal{C}}(\mathcal{C}) f_\alpha^{\tilde{c}}(\tilde{c}_\alpha). \tag{4.81}$$

Thereby, S denotes the ASTM austenite grain size number $\tilde{c}_\alpha$ the phantom volume fraction of phase α, and $\mathcal{C} = \{\text{C,Mn,B,Si,}\dots\}$ the set of chemical components of the considered steel. The normalized volume fraction must be corrected to consider the thermodynamic equilibrium state of ferrite and pearlite for a given temperature. The ferritic normalized volume fraction phase has to be corrected via $c_f = \tilde{c}_f c_f^{eq}$, where c_f^{eq} is the equilibrium ferrite volume fraction. For pearlite, the relation reads $c_p = \tilde{c}_p(1 - c_f^{eq})$ (see, e.g. Bok et al., 2011). The equilibrium ferrite volume fraction, which is dependent on temperature and chemical composition, is obtained from the corresponding Fe3C diagram or by empirical relations from Andrews (1965). For bainite, the normalized fraction corresponds to the volume fraction.

The influence function of the austenitic grain size f_α^S, of the temperature f_α^θ, the chemical composition $f_\alpha^{\mathcal{C}}$, and the normalized volume fraction $f_\alpha^{\tilde{c}}$ has been calibrated with various TTT curves in the US Steel Atlas (see, e.g., United States Steel Company Research Laboratory, 1963) by Kirkaldy and Venugopalan (1983). To improve the estimation of the hardness of a metal after heat treatment, Li et al. (1998) modified the model by calibrating the KV model to CCT diagrams. For the steel 22MnB5, Ackerström and Oldenburg (2006) extended the reaction rate of Li's model by considering the influence of boron, resulting in the influence functions (see, e.g., Bok et al., 2011)

$$f_f^S(S) = 2^{0.41S}, \qquad f_f^\theta(\theta) = (Ae_3 - \theta)^3 \exp\left(-\frac{A_f}{R\theta}\right), \tag{4.82}$$

$$f_f^C(C) = 59.6C_{\text{Mn}} + 1.45C_{\text{Ni}} + 67.6C_{\text{Cr}} + 244C_{\text{Mo}} + 1.9 \cdot 10^5 C_{\text{B}} \tag{4.83}$$

for ferrite, with the equilibrium temperature Ae_3, which is the upper limit of the regime at which ferrite and austenite can coexist (see, e.g., Ali and Bhadeshia, 1990),

$$f_p^S(S) = 2^{0.32S}, \qquad f_p^\theta(\theta) = (Ae_1 - \theta)^3 \exp\left(-\frac{A_p}{R\theta}\right) \tag{4.84}$$

$$f_p^C(C) = 1.79 + 5.42(C_{\text{Cr}} + C_{\text{Mo}} + 4C_{\text{Mo}}C_{\text{Ni}}) + 3.1 \cdot 10^3 C_{\text{B}} \tag{4.85}$$

for pearlite, with the equilibrium temperature Ae_1, which is the lower limit of the regime at which ferrite and pearlite can coexist (see, e.g., Equihua and Salinas, 2010), and

$$f_b^S(S) = 2^{0.29S}, \qquad f_f^\theta(\theta) = (Bs - \theta)^3 \exp\left(-\frac{A_b}{R\theta}\right), \tag{4.86}$$

$$f_f^C(C) = 59.6C_{\text{Mn}} + 1.45C_{\text{Ni}} + 67.6C_{\text{Cr}} + 244C_{\text{Mo}} + 1.9 \cdot 10^5 C_{\text{B}}, \tag{4.87}$$

$$f_b^{\tilde{c},C}(\tilde{c}_b) = \frac{\tilde{c}_b^{0.4(1-\tilde{c}_b)} (1 - \tilde{c}_b)^{0.4\tilde{c}_b}}{\exp\left(\tilde{c}_b^2(1.9C_{\text{C}} + 2.5C_{\text{Mn}} + 9C_{\text{Ni}} + 1.7C_{\text{Cr}} + 4C_{\text{Mo}} - 2.6)\right)} \tag{4.88}$$

for bainite, with the bainite formation start temperature Bs. Thereby, the universal gas constant is denoted by R. Note, that, in this approach, the influence function of the normalized volume fraction and the influence function of the chemical composition are not uncoupled. The activation energy for the phase transformation of the different phases is denoted by A_α . For ferrite and pearlite, the influence function of the normalized volume fraction reads

$$f_{f,p}^{\tilde{c}}(\tilde{c}_{f,p}) = \tilde{c}_{f,p}^{0.4(1-\tilde{c}_{f,p})} (1 - \tilde{c}_{f,p})^{0.4\tilde{c}_{f,p}} . \tag{4.89}$$

If empirical relations for the equilibrium temperatures and the formation start temperatures are introduced dependent on the chemical composition (see, e.g., Kirkaldy and Venugopalan, 1983; Lee et al., 2009)

$$\begin{aligned} Ae3 &= (912 - 203\sqrt{C_{\text{C}}} - 15.2C_{\text{Ni}} + 44.7C_{\text{Si}} + 31.5C_{\text{Mo}} - 30C_{\text{Mn}} - \\ &\qquad 11C_{\text{Cr}} + 700C_{\text{P}} + 400C_{\text{Al}} + 400C_{\text{Ti}}) + 273.15, \qquad (4.90) \\ Ae1 &= (723 - 10.7C_{\text{Mn}} - 16.9C_{\text{Ni}} + 29C_{\text{Si}} + 16.9C_{\text{Cr}}) + 273.15, \qquad (4.91) \\ Bs &= (656 - 58C_{\text{C}} - 35C_{\text{Mn}} - 75C_{\text{Si}} - 15C_{\text{Ni}} - 34C_{\text{Cr}} - \\ &\qquad 41C_{\text{Mo}}) + 273.15, \qquad (4.92) \\ Ms &= (561 - 474C_{\text{C}} - 33C_{\text{Mn}} - 17C_{\text{Ni}} - 17C_{\text{Cr}} - 21C_{\text{Mo}}) + 273.15, \qquad (4.93) \end{aligned}$$

where Ms denotes the martensite formation start temperature, the KV model does not need any experimental results, besides the chemical composition and the grain size of the austenitic phase, to predict the diffusion-driven transformation behavior of an arbitrary steel.

In this work, a simplified KV model is suggested. Since ferrite and pearlite are assumed to be one phase, the normalized phase and the relation for the equilibrium ferrite volume fraction are not needed. Inspired by the temperature dependence of the JMAK parameter (4.77), the influence function for the temperature

$$f_{\alpha}^{\theta}(\theta) = \exp\left(\frac{-\frac{A_{\alpha}}{(Ae3-\theta)^2 + d_{\alpha}}}{R\theta} \right) \tag{4.94}$$

is suggested. The other influence functions read

$$f_{\alpha}^{S}(S) = 2^{m_{\alpha}S}, \quad f_{\alpha}^{c}(c_{\alpha}) = c_{\alpha}^{n_{\alpha}(1-c_{\alpha})} (1 - c_{\alpha})^{n_{\alpha}c_{\alpha}} . \tag{4.95}$$

For the influence function of the chemical composition of the ferritic/pearlitic and bainitic phase, the standard influence function of ferrite (4.83) and bainite (4.87), respectively, are chosen. In contrast to Li's model, in this modification, the proposed simplification yields a multiplicative decomposition of the influences for all possible phase transformation.

Non-isothermal Phase Transformation

The JMAK equation and the KV model are originally derived under the assumption of an isothermal heat treatment. To be able to use these models in the non-isothermal case, Scheil's additivity rule is used (Scheil, 1935). Scheil derived the rule to predict the incubation time arising in diffusion-driven phase transformation. The idea is to decompose the thermomechanical non-isothermal process in n small isothermal steps of time length $\Delta t = t_p/n$, where t_p is the overall process time. For a given volume fraction $c_\alpha(t)$ of phase α, the time that is needed to get this volume fraction under isothermal condition at the temperature level of the k-th step $\tau_\alpha(\theta(k\Delta t), c_\alpha(t))$ is determinable. The sum of the time fractions defined by the time available for the transformation Δt by the time needed for the transformation τ_α at the temperature level $\theta(k\Delta t)$ is one. In the limit $n \to \infty$, Scheil's additivity rule

$$\int_0^t \frac{1}{\tau_\alpha(\theta(\xi), c_\alpha(t))} \, \mathrm{d}\xi = 1 \tag{4.96}$$

is obtained. Lusk and Jou (1997), Todinov (1998), and Rios (2005) for example, have discussed the applicability of evolution laws in the context of Scheil's additivity rule. An evolution equation is called additive if it fulfills Scheil's additivity rule. Lusk and Jou (1997) found that an evolution equation is additive only if the constitutive law is iso-kinetic, i.e. if it is decomposable in a multiplicative way into an influence function of the volume fraction and an influence function of the temperature

$$\dot{c}_\alpha = g^c_\alpha(c_\alpha)g^\theta_\alpha(\theta). \tag{4.97}$$

For the JMAK equation, this is the case only if the JMAK exponent a_α is constant. The derivation of equation (4.75) with respect to time yields the rate law of the JMAK equation

$$\dot{c}_\alpha = (c_\alpha - c^r_\alpha)\ln\left(1 - \frac{c_\alpha}{c^r_\alpha}\right)^{\frac{a_\alpha - 1}{a_\alpha}} a_\alpha b_\alpha(\theta)^{\frac{1}{a_\alpha}}, \tag{4.98}$$

which shows that the model is iso-kinetic. Lusk and Jou (1997) showed that the deviation of the CCT diagram resulting from a JMAK model with a non constant Avrami exponent the CCT diagram resulting from an additive JMAK model is slight.

Due to its multiplicative nature, the KV evolution law (4.81) is directly iso-kinetic and, thus, additive.

Using the JMAK equation, the time needed to get the volume fraction c_α under isothermal condition is obtained by rearranging of equation (4.75)

$$\tau_\alpha\left(\theta\left(t\right), c_\alpha\left(t\right)\right) = \left(-\frac{\ln\left(1 - \frac{c_\alpha(t)}{c^r_\alpha}\right)}{b_\alpha\left(\theta\right)}\right)^{\frac{1}{a_\alpha(\theta)}}. \tag{4.99}$$

Deriving equation (4.96) with respect to time t, Hömberg (1996) derived another rate law for the evolution of the volume fraction of phase α

$$\dot{c}_\alpha\left(t\right) = -\left(\int_0^t -\frac{\tau_\alpha(\theta(\xi), c_\alpha(t))^{-1-a_\alpha(\theta(\xi))}}{a_\alpha(\theta(\xi))b_\alpha(\theta(\xi))\left(1 - c_\alpha(t)\right)}\,\mathrm{d}\xi\right)^{-1}\frac{1}{\tau_\alpha(\theta(t), c_\alpha(t))}. \tag{4.100}$$

It is observed, that the JMAK equation and the KV model, whose parameters are fitted to the TTT diagram of 22MnB5, deviate from experimental results if continuous cooling processes are considered. To overcome such discrepancies, several authors, e.g., Kamamoto et al.

(1985), Boyadjiev et al. (1996), and Bok et al. (2015), have proposed different modifications of the transformation time or JMAK parameters to take the cooling rate into account. In this work, a multiplicative decomposition of the JMAK parameters into a the standard isothermal part and a cooling rate depending part

$$a_\alpha(\theta,\dot{\theta}) = a^\theta_\alpha(\theta)a^r_\alpha(\dot{\theta}), \qquad b_\alpha(\theta,\dot{\theta}) = b^\theta_\alpha(\theta)b^r_\alpha(\dot{\theta}) \tag{4.101}$$

is proposed. In the non-isothermal case, the JMAK exponent is modified with a constant factor and the JMAK factor with a quadratic function beyond a critical cooling rate

$$a^r_\alpha = 1 + (a^r_{0,\alpha} - 1)\mathcal{H}(-\dot{\theta}) \tag{4.102}$$

$$b^r_\alpha = \begin{cases} 1 + (b^r_{0,\alpha} - 1)\mathcal{H}(-\dot{\theta}) & \text{if} \quad \dot{\theta} \leq b^r_{1,\alpha} \\ 1 + \left(\frac{1-b^r_{0,\alpha}}{b^r_{1,\alpha}(2b^r_{2,\alpha}-b^r_{1,\alpha})}\dot{\theta}^2 + 2\frac{b^r_{2,\alpha}(b^r_{0,\alpha}-1)}{b^r_{1,\alpha}(2b^r_{2,\alpha}-b^r_{1,\alpha})}\dot{\theta}\right)\mathcal{H}(-\dot{\theta}) & \text{else.} \end{cases} \tag{4.103}$$

The KV model is modified by adding an influence function of the cooling rate $f^r_\alpha(\dot{\theta})$ to equation (4.81)

$$\dot{c}_\alpha = \dot{c}_\alpha(\tilde{c}_\alpha, \theta, \dot{\theta}, S, C) = f^S_\alpha(S)f^\theta_\alpha(\theta)f^C_\alpha(C)f^c_\alpha(c_\alpha)f^r_\alpha(\dot{\theta}), \tag{4.104}$$

which, similar to the JMAK extension, is chosen to be a quadratic function of the cooling rate

$$f^r_\alpha(\dot{\theta}) = 1 + \left(\frac{1 - r_{0,\alpha}}{r_{1,\alpha}(2r_{2,\alpha} - r_{1,\alpha})}\dot{\theta}^2 + 2\frac{r_{2,\alpha}(r_{0,\alpha} - 1)}{r_{1,\alpha}(2r_{2,\alpha} - r_{1,\alpha})}\dot{\theta}\right)\mathcal{H}(-\dot{\theta}). \tag{4.105}$$

4.4.2 The Diffusionless Phase Transformation

The diffusionless phase transformation of martensite from austenite is usually modeled by the Koistinen-Marburger law. In experiments, Koistinen and Marburger (1959) observed a proportional connection between the logarithm of the austenitic phase volume fraction and the undercooling below the martensite formation start temperature

$$\ln\left(c_a\right) = \gamma\left(M_s - \theta\right). \tag{4.106}$$

For temperatures beyond Ms, martensitic transformation is not possible. Thereby, Wang et al. (2000) showed that the MS is dependent on the carbon concentration in the austenitic phase. Todinov et al. (1996) and Alexander et al. (1995) investigated the stress dependence of the MS. The carbon concentration in the austenitic phase is influenced by prior diffusion-driven phase transformation which leads to a decrease of the MS. In this work, both effects are not considered.

Taking previous phase transformations into account, the rearranging of equation (4.106) yields the so called Koistinen-Marburger (KM) model valid for isothermal processes (see, e.g., Neumann and Böhlke, 2016)

$$c_m = c_m^r\left(1 - \exp\left(\gamma\left(M_s - \theta\right)\right)\right), \tag{4.107}$$

where γ denotes the transition parameter. In dilation experiments with some steels, in the cooling part of the dilatation curve, one observes that the martensitic transformation is not just an exponential-like behavior predicted by the KM model but follows an "S"-shaped curve. To capture this behavior, Bok et al. (2014) proposed a new model by combining the original KM model with an additional exponential curve. A critical temperature is introduced. Below this critical temperature the KM model is valid, whereas beyond this temperature, the additional curve is valid leading to the desired "S"-shape transformation behavior. Due to the discontinuity of the first derivation with respect to the temperature of Bok's model at the critical temperature, a sigmoidal approach is suggested

$$c_m = \frac{c_m^r}{1 + a\exp\left(b\left(\theta - c\right)\right)} \tag{4.108}$$

leading to a continuous function even in the first derivation with respect to temperature.

The non-isothermal transformation from austenite into martensite is usually modeled in two ways: the first is to use equation (4.107) directly (see, e.g., Ackerström and Oldenburg, 2006; Olsson, 2009; Naderi, 2007) whereas the second way is to introduce a rate law describing the kinetics of the martensitic transformation (Hömberg, 1996). In this work, following De Oliveira et al. (2010), a rate law consistent to the KM model is used

$$\dot{c}_m = \gamma\dot{\theta}\left(c_m^r - c_m\right)\mathcal{H}\left(-\dot{\theta}\right)\mathcal{H}\left(M_s - \theta\right). \tag{4.109}$$

The rate law consistent to the sigmoidal approach reads

$$\dot{c}_m = b\dot{\theta}\left(c_m - c_m^r\right)\frac{c_m}{c_m^r}\mathcal{H}\left(-\dot{\theta}\right)\mathcal{H}\left(M_s - \theta\right), \tag{4.110}$$

which appears to be a nonlinear extension of the KM rate law (4.109).

Chapter 5

The Scale Transition Problem - Homogenization and Localization

5.1 A Phenomenological Reference Model

A phenomenological model is now introduced which is similar to the models usually used in the context of simulation of the hot stamping process. This model serves as a reference model for the thermo-micromechanical (TMM) model developed in this work.

In the following, quantities marked with a bar refer to the macro scale. The phase volume average of an arbitrary quantity is given by equation (3.38) and is related to the micro scale.

The macroscopic reference model is based on the macroscopic Helmholtz free energy corresponding to (4.11)

$$\bar{\rho}\bar{\psi} = \frac{1}{2}\left(\bar{\varepsilon} - \bar{\varepsilon}^E\right) \cdot \bar{\mathbb{C}}\left[\bar{\varepsilon} - \bar{\varepsilon}^E\right] - \frac{1}{2}\bar{\alpha} \cdot \bar{\mathbb{C}}\left[\bar{\alpha}\right]\Delta\bar{\theta}^2 + \bar{\rho}\bar{\kappa}_c\left(\Delta\bar{\theta} - \bar{\theta}\ln\left(\frac{\bar{\theta}}{\theta_0}\right)\right) + \bar{\rho}\bar{\psi}^i\left(\bar{\varepsilon}^p\right), \quad (5.1)$$

leading to the stress-strain relation

$$\bar{\sigma} = \bar{\mathbb{C}}\left[\bar{\varepsilon} - \bar{\varepsilon}^p - \bar{\varepsilon}^{tr} - \bar{\varepsilon}^t - \bar{\alpha}\Delta\bar{\theta}\boldsymbol{I}\right] \quad (5.2)$$

derived via the CD inequality on the macro scale and the resulting potential relations on the micro scale corresponding to (3.34) and (4.4), respectively. The stiffness tensor is determined by a simple mixture rule reading

$$\bar{\mathbb{C}} = 3\,\langle K\rangle\,\mathbb{P}^{1,\,\mathrm{iso}} + 2\,\langle G\rangle\,\mathbb{P}^{2,\,\mathrm{iso}} \tag{5.3}$$

for isotropic elastic phases. Furthermore, the additive decomposition of the macroscopic strain tensor

$$\bar{\varepsilon} = \bar{\varepsilon}^{e} + \bar{\varepsilon}^{p} + \bar{\varepsilon}^{tr} + \bar{\varepsilon}^{t} + \bar{\alpha}\Delta\bar{\theta}\boldsymbol{I} \tag{5.4}$$

and an isotropic macroscopic thermal expansion have been assumed which is determined by a simple mixture rule $\bar{\alpha} = \langle\alpha\rangle$. The macroscopic plastic strain, macroscopic TRIP strain, and the macroscopic transformation strain are modeled via the principle of maximum plastic dissipation

$$\dot{\bar{\varepsilon}}^{p} = \dot{\bar{\gamma}}\frac{\bar{\boldsymbol{\sigma}}'}{\|\bar{\boldsymbol{\sigma}}'\|}, \qquad \dot{\bar{\varepsilon}}^{p} = \dot{\bar{\gamma}}\sqrt{\frac{2}{3}}, \tag{5.5}$$

via Leblond's equation corresponding to equation (4.50) with the extension by Wolff et al. (2008b) and De Oliveira et al. (2010)

$$\dot{\bar{\varepsilon}}^{tr} = \frac{3}{2}\sum_{\alpha\in\mathcal{PH}} k_{\alpha}^{tr}\varphi_{\alpha}'(c_{\alpha})\dot{c}_{\alpha}\mathcal{H}(\dot{c}_{\alpha})\bar{\boldsymbol{\sigma}}', \tag{5.6}$$

and isotropic transformation strain (cf. equation (4.57))

$$\bar{\varepsilon}^{t} = \langle\zeta\mathcal{H}\left(c_{\alpha}\right)\rangle\,\boldsymbol{I}, \tag{5.7}$$

respectively. Thereby, the macroscopic yield surface reads (cf. equation (4.39))

$$\bar{\phi}\left(\bar{\boldsymbol{\sigma}},\bar{\varepsilon}^{p}\right) = \|\bar{\boldsymbol{\sigma}}'\| - \sqrt{\frac{2}{3}}\left(\bar{\sigma}^{0} + \bar{\iota}^{p}\right), \tag{5.8}$$

where the macroscopic initial yield stress is the average of the initial yield stress of the existing phases $\bar{\sigma}_0 = \langle \sigma_0 \rangle$. The thermodynamic conjugate of the macroscopic accumulated plastic strain $\bar{\iota}^p$ is given by

$$\bar{\iota}^p = \bar{\rho} \frac{\partial \bar{\psi}^i}{\partial \bar{\varepsilon}^p}. \tag{5.9}$$

This approach takes the different plastic behaviors of the phase by $\bar{\rho} \bar{\psi}^i = \langle \rho \psi^i \rangle$ into account, leading to the macroscopic yield surface

$$\bar{\phi}\left(\bar{\boldsymbol{\sigma}}, \bar{\varepsilon}^p\right) = \|\bar{\boldsymbol{\sigma}}'\| - \sqrt{\frac{2}{3}} \left\langle s^f\left(\bar{\theta}\right)\left(\varepsilon^0 + \bar{\varepsilon}^p\right)^{s^e} \right\rangle, \tag{5.10}$$

if Swift's hardening potential for each phase

$$\rho_\alpha \psi_\alpha^i\left(\bar{\varepsilon}^p\right) = \frac{s_\alpha^f(\bar{\theta})}{s_\alpha^e(\bar{\theta}) + 1}\left(\varepsilon_\alpha^0 + \bar{\varepsilon}^p\right)^{s_\alpha^e + 1} - s_\alpha^f(\bar{\theta})\left(\varepsilon_\alpha^0\right)^{s_\alpha^e} \bar{\varepsilon}^p \tag{5.11}$$

is used (cf. equation (4.47)).

In the special case of a process with a constant macroscopic stress, an interesting interpretation of the TRIP effect can be found. In such a case, the TRIP strain (5.6) can be directly integrated leading to the deviatoric macroscopic stress

$$\bar{\boldsymbol{\sigma}}' = \frac{2\langle G \rangle}{\gamma^{tr}}\left(\bar{\boldsymbol{\varepsilon}}' - \bar{\boldsymbol{\varepsilon}}^p\right) \tag{5.12}$$

with the shear softening parameter

$$\gamma^{tr} = 1 + \frac{3}{2} \sum_{\alpha \in \mathcal{PH}} k_\alpha^{tr} \phi(c_\alpha) \mathcal{H}(\Delta c_\alpha). \tag{5.13}$$

Since $\gamma^{tr,n+1} \geq 1$ holds, the TRIP softens the elastic shear behavior of the bulk.

The temperature development is described by a macroscopic heat equation corresponding to equation (4.10)

$$\bar{\rho}\bar{\kappa}\dot{\bar{\theta}} - \operatorname{div}\left(\bar{\boldsymbol{\lambda}}\operatorname{grad}\left(\bar{\theta}\right)\right) = \sum_{\alpha\in\mathcal{PH}\setminus 1} L_\alpha \dot{c}_\alpha, \tag{5.14}$$

where the mechanical dissipation contributions have been neglected and the energy sources and sinks are given by the latent heat L_α released or consumed during phase transformation. Without loss of generality, the initial phase, which is decomposed, is called 1. The macroscopic specific heat capacity and the macroscopic thermal conductivity are given by the mixture rule

$$\bar{\kappa} = \langle\kappa\rangle, \qquad \bar{\boldsymbol{\lambda}} = \langle\lambda\rangle\, \boldsymbol{I}, \tag{5.15}$$

respectively, where the phases' thermal conductivity is assumed to be isotropic.

For the description of the phase transformation, the JMAK or KV model are used in the diffusion-driven case and the KM model in the diffusionless case.

5.2 Thermomechanical Behavior of Phases for the Two-Scale Model[1]

In a mean field theory, the thermomechanical material behavior is determined by phase-wise constant thermomechanical quantities, cf. equation (3.36). With a phase-wise constant elastic strain, eigenstrain, and thermomechanical properties, the Helmholtz free energy on the mi-

[1] The modeling approach for the phase behavior, which is described in this chapter, is based on the paper "Hashine-Shtrikman type mean field model for the two-scale simulation of the thermomechanical processing of steel" (Neumann and Böhlke, 2016) and is extended by a model which takes the TRIP-effect into account.

crostructure is a phase-wise constant itself and can be expressed by

$$\psi\left(\boldsymbol{x}\right) = \sum_{\alpha\in\mathcal{PH}} \chi_{\alpha}\left(\boldsymbol{x}\right)\psi_{\alpha}. \tag{5.16}$$

Assuming a phases' Helmholtz free energy corresponding to equation (4.11)

$$\rho_{\alpha}\psi_{\alpha} = \frac{1}{2}\left(\boldsymbol{\varepsilon}_{\alpha} - \boldsymbol{\varepsilon}_{\alpha}^{E}\right)\cdot\mathbb{C}_{\alpha}\left[\boldsymbol{\varepsilon}_{\alpha} - \boldsymbol{\varepsilon}_{\alpha}^{E}\right] - \frac{1}{2}\boldsymbol{\alpha}_{\alpha}\cdot\mathbb{C}_{\alpha}\left[\boldsymbol{\alpha}_{\alpha}\right]\Delta\theta^{2} + \rho_{\alpha}\kappa_{\alpha}^{c}\left(\Delta\theta - \theta\ln\left(\frac{\theta}{\theta_{0}}\right)\right) + \rho_{\alpha}\psi_{\alpha}^{i}\left(\varepsilon_{\alpha}^{p}\right), \tag{5.17}$$

the Hooke's law

$$\boldsymbol{\sigma}_{\alpha} = \mathbb{C}_{\alpha}\left[\boldsymbol{\varepsilon}_{\alpha} - \boldsymbol{\varepsilon}_{\alpha}^{E}\right], \tag{5.18}$$

is obtained by use of potential relations equivalent to (4.4). The elasticity of the phases is assumed to be isotropic

$$\mathbb{C}_{\alpha} = 3K_{\alpha}\mathbb{P}^{1,\,\mathrm{iso}} + 2G_{\alpha}\mathbb{P}^{2,\,\mathrm{iso}}, \tag{5.19}$$

where the temperature dependence of the stiffness tensor is captured by the temperature dependence of the Young's modulus and Poisson's number. They are assumed to be linear

$$E_{\alpha}(\theta) = m_{\alpha}^{E}\theta + n_{\alpha}^{E}, \quad \nu_{\alpha}(\theta) = m_{\alpha}^{\nu}\theta + n_{\alpha}^{\nu}. \tag{5.20}$$

Assuming an isotropic temperature expansion coefficient, the eigenstrain tensor is given by

$$\boldsymbol{\varepsilon}_{\alpha}^{E} = \boldsymbol{\varepsilon}_{\alpha}^{p} + \boldsymbol{\varepsilon}_{\alpha}^{tr} + \boldsymbol{\varepsilon}_{\alpha}^{t} + \alpha_{\alpha}\Delta\theta\boldsymbol{I}. \tag{5.21}$$

In a maximum dissipation framework, the evolution equations for the phases' plastic strain and equivalent plastic strain are given by

$$\dot{\boldsymbol{\varepsilon}}^p_\alpha = \dot{\gamma}_\alpha \frac{\boldsymbol{\sigma}'_\alpha}{\|\boldsymbol{\sigma}'_\alpha\|}, \qquad \dot{\varepsilon}^p_\alpha = \dot{\gamma}_\alpha \sqrt{\frac{2}{3}}. \tag{5.22}$$

With the hardening potentials

$$\rho_\alpha \psi^i_\alpha \left(\varepsilon^p_\alpha\right) = \frac{1}{2} \Theta^\infty_\alpha \, \varepsilon^{p\,2}_\alpha + \left(\sigma^\infty_\alpha - \sigma^0_\alpha\right) \Bigg(\varepsilon^p_\alpha - \frac{\sigma^\infty_\alpha - \sigma^0_\alpha}{\Theta^\infty_\alpha - \Theta^0_\alpha} \exp\left(\frac{\Theta^\infty_\alpha - \Theta^0_\alpha}{\sigma^\infty_\alpha - \sigma^0_\alpha} \varepsilon^p_\alpha \right) \Bigg), \tag{5.23}$$

for the Voce-type hardening, and

$$\rho_\alpha \psi^i_\alpha \left(\varepsilon^p_\alpha\right) = \frac{s^f_\alpha(\theta)}{s^e_\alpha(\theta) + 1} \left(\varepsilon^0_\alpha + \varepsilon^p_\alpha\right)^{s^e_\alpha + 1} - s^f_\alpha(\theta) \left(\varepsilon^0_\alpha\right)^{s^e_\alpha} \varepsilon^p_\alpha, \tag{5.24}$$

for the Swift-type hardening, the associated yield surfaces

$$\phi_\alpha = \|\boldsymbol{\sigma}'_\alpha\| - \sqrt{\frac{2}{3}} \Bigg(\sigma^0_\alpha + \Theta^\infty_\alpha \varepsilon^p_\alpha + \left(\sigma^\infty_\alpha - \sigma^0_\alpha\right) \left(1 - \exp\left(\frac{\Theta^\infty_\alpha - \Theta^0_\alpha}{\sigma^\infty_\alpha - \sigma^0_\alpha} \varepsilon^p_\alpha \right) \right) \Bigg) \leq 0 \tag{5.25}$$

and

$$\phi_\alpha = \|\boldsymbol{\sigma}'_\alpha\| - \sqrt{\frac{2}{3}} s^f_\alpha(\theta) \left(\varepsilon^0_\alpha + \varepsilon^p_\alpha\right)^{s^e_\alpha} \leq 0 \tag{5.26}$$

are obtained, respectively. Thereby, the temperature-dependent Swift parameters are given by

$$s^f_\alpha(\theta) = s^{f,0}_\alpha + s^{f,1}_\alpha \theta, \qquad s^e_\alpha(\theta) = s^{e,0}_\alpha + s^{e,1}_\alpha \theta. \tag{5.27}$$

Plastic yield occurs if and only if the yield condition and the loading condition

$$\phi_\alpha = 0, \qquad \frac{\partial \phi_\alpha}{\partial \boldsymbol{\varepsilon}^e_\alpha} \cdot \dot{\boldsymbol{\varepsilon}}^e_\alpha + \frac{\partial \phi_\alpha}{\partial \theta}\dot{\theta} > 0. \tag{5.28}$$

are fulfilled.

Since the macroscopic TRIP strain is an accumulation of the micro-plastic strains in the weaker phase caused by the volume change of the growing phase during phase transformation, only the parent phase experiences a TRIP strain. According to Leblond's approach and to extensions of, e.g., Wolff et al. (2008b) or De Oliveira et al. (2010), the TRIP strain in the parent phase, i.e. the austenitic phase in the hot stamping process, can be modeled by

$$\dot{\boldsymbol{\varepsilon}}^{tr}_a = \frac{3}{2} \sum_{\alpha \in \mathcal{PH} \backslash a} k^{tr}_\alpha \varphi'_\alpha(c_\alpha)\dot{c}_\alpha \bar{\boldsymbol{\sigma}}' \mathcal{H}(\dot{c}_\alpha), \tag{5.29}$$

with the saturation functions given in (4.51), (4.52), and (4.53).

It is observed that, since all the information of the TRIP is stored in the austenitic microscopic TRIP strain, in such a model, the macroscopic TRIP strain is lost when the entire austenite is transformed. This modeling approach is based on the assumption, that the dislocations vanish after the transformation front of the growing phases and no plasticity is transferred from the parent phase into the growth phase. Since this assumption leads to unphysical results for this model, the TRIP in the parent phase has to be transferred to the growing phase. An initial approach is

$$\dot{\boldsymbol{\varepsilon}}^{tr}_\alpha = \dot{\boldsymbol{\varepsilon}}^{tr}_a \mathcal{H}(\dot{c}_\alpha) \quad \forall \alpha \in \mathcal{PH} \backslash a, \tag{5.30}$$

taking the entire TRIP of the parent phase into account in the growth phase.

Note that Tang et al. (2014) found in experimental investigations, that the TRIP parameter k^{tr}_α is dependent on the applied stress. In this work, a linear

$$k_{\alpha}^{tr} = k_{0,\alpha}^{l} + k_{1,\alpha}^{l} \tag{5.31}$$

and a Weibull's function approach

$$k_{\alpha}^{tr} = k_{0,\alpha}^{w} - k_{1,\alpha}^{w} \exp\left(-k_{2,\alpha}^{w} \left(\sigma_{\alpha}^{v}\right)^{k_{3,\alpha}^{w}}\right) \tag{5.32}$$

are considered to describe the applied stress dependence. Thereby, the v. Mises stress

$$\sigma_{\alpha}^{v} = \sqrt{\frac{3}{2}\boldsymbol{\sigma}'_{\alpha} \cdot \boldsymbol{\sigma}'_{\alpha}} \tag{5.33}$$

has been introduced.

Note that, in a mean field approach, forces on the interface between the parent phase and the growing phase are not taken into account. In such an approximation, volume changes due to transformation or thermal strain of a growing phase, cause only spherical strains in the parent phase and vice versa. Since the plasticity is a deviatoric effect, no plasticity due to different thermal expansions of the parent and growing phase can occur in the parent phase, which is the determining factor of the TRIP effect after Leblond et al. (1989). For this reason, the microscopic TRIP strain tensor has to be introduced to capture the TRIP effect.

As discussed in Section 4.3, the transformation strain is determined by the transformation of the parent phase into a microstructure consisting of different variants of the new phase. Each variant v of the new phase α is assigned with the corresponding deviatoric shape change of the lattice $\boldsymbol{\Xi}_{v}^{\alpha}$. Adding all transformation strains of the variants, which contributes to the total transformation strain of the growing phase, leads to

$$\boldsymbol{\varepsilon}_{\alpha}^{t} = \sum_{v=1}^{N_{v}^{\alpha}} c_{v}^{\alpha} \boldsymbol{\Xi}_{v}^{\alpha}, \tag{5.34}$$

where N_{v}^{α} denotes the number of variants and c_{v}^{α} the volume fraction of variant v of phase α fulfilling $\sum_{v=1}^{N_{v}^{\alpha}} c_{v}^{\alpha} = c_{\alpha}$. With an uniform distribution of the variants $c_v^\alpha = c_\alpha / N_v^\alpha$, equivalent to (4.57), the phases' transformation strain can be assumed to be isotropic

$$\varepsilon_\alpha^t = \zeta_\alpha \mathcal{H}(c_\alpha)\boldsymbol{I} \tag{5.35}$$

with the mean transformation volume change ζ_α.

Note, that with the concept of the mean instantaneous transformation strain (MITS) and the Sachs approximation (see Appendix B.2), Kubler et al. (2011) derived a trip evolution equation taking the orientation effect into account. In this approach, the transformation strain and Leblond's TRIP strain, as well, are included as a special case.

5.3 Thermomechanical Two-Scale Problem

In a mean field model, the interaction between the scales is described by localization and homogenization relations. In thermomechanical processes, both the thermal-related quantities, heat flux and the temperature gradient, and the mechanical-related quantities, stress and strain, on the micro scale have to be related to the corresponding quantities on the macro scale. A set of estimates and bounds for the thermal and mechanical behavior has been proposed in literature briefly outlined in Appendix B. Thereby, due to similarity of the determining BVPs, the thermal problem and the linear mechanical problem can be treated in an equivalent way (see, e.g., the energetic first-order bounds for both the thermal and mechanical problems in Appendix B.2). However, while, in the thermal case, due to a linear constitutive relation between heat flux and temperature gradient, the thermal localization relation is linear, in the mechanical case, due to nonlinear plastic and TRIP strain, the mechanical localization relation is nonlinear. For the Hashin-Shtrikman-type (HS) homogenization schemes, for example, this means that the thermal localization is of second-order while it is not evident if the mechanical localization is of second-order.

5.3.1 Hashin-Shtrikman Estimate of Thermal Properties

Hashin and Shtrikman (1962b) suggested a variational homogenization method leading to second-order bounds for the thermal or mechanical effective properties in the linear case. The method is based on the variation of the difference between the energy stored in an introduced homogeneous comparison material and the stored energy in the inhomogeneous microstructure. In the thermal case, this difference, also referred to as HS functional, reads

$$\mathcal{F}^{\theta} = W_0^{\theta} - \bar{W}^{\theta} = \frac{1}{2}\bar{\boldsymbol{g}} \cdot \boldsymbol{\lambda}_0 \bar{\boldsymbol{g}} - \left\langle \frac{1}{2}\boldsymbol{g} \cdot \boldsymbol{\lambda} \boldsymbol{g} \right\rangle, \tag{5.36}$$

where $\boldsymbol{\lambda}_0$ denotes the thermal conductivity tensor of the homogeneous comparison material. Assuming a/an

(i) Constant volume fractions of the phases,
(ii) Phase-wise constant heat flux polarization,
(iii) Statistical homogeneity and no long-range order,
(iv) Isotropic comparison material,
(v) Isotropic two point statistics,

and isotropic phases' thermal conductivity, the localization relation for the temperature gradient

$$\boldsymbol{g}_{\alpha} = \frac{\left\langle \frac{1}{1+\frac{\lambda_{\alpha}-\lambda_0}{3\lambda_0}} \right\rangle^{-1}}{1 + \frac{\lambda_{\alpha}-\lambda_0}{3\lambda_0}} \bar{\boldsymbol{g}} = \frac{\left\langle \frac{1}{\lambda_{\alpha}+2\lambda_0} \right\rangle^{-1}}{\lambda_{\alpha} + 2\lambda_0} \bar{\boldsymbol{g}} \tag{5.37}$$

can be derived from the HS functional. The first-order upper and lower bound (cf. Appendix B.2) are included for $\lambda_0 \to \infty$ and $\lambda_0 \to 0$, respectively. Two effective thermal conductivities can be derived by using the constitutive law

$$\bar{\boldsymbol{\lambda}}^{q} = \left\langle \frac{\lambda}{1+\frac{\lambda-\lambda_0}{3\lambda_0}} \right\rangle \left\langle \frac{1}{1+\frac{\lambda-\lambda_0}{3\lambda_0}} \right\rangle^{-1} \boldsymbol{I} = \left\langle \frac{\lambda}{\lambda+2\lambda_0} \right\rangle \left\langle \frac{1}{\lambda+2\lambda_0} \right\rangle^{-1} \boldsymbol{I} \tag{5.38}$$

or the stored thermal energy

$$\begin{aligned} \bar{\boldsymbol{\lambda}}^{W} &= \left\langle \frac{\lambda}{\left(1+\frac{\lambda-\lambda_0}{3\lambda_0}\right)^2} \right\rangle \left\langle \frac{1}{1+\frac{\lambda-\lambda_0}{3\lambda_0}} \right\rangle^{-2} \boldsymbol{I} \\ &= \left\langle \frac{\lambda}{(\lambda+2\lambda_0)^2} \right\rangle \left\langle \frac{1}{\lambda+2\lambda_0} \right\rangle^{-2} \boldsymbol{I}. \end{aligned} \tag{5.39}$$

Note, that, if and only if the thermal Hill-Condition $\langle \boldsymbol{q} \cdot \boldsymbol{g} \rangle = \langle \boldsymbol{q} \rangle \cdot \langle \boldsymbol{g} \rangle$ (see, e.g., Ostoja-Starzewski, 2005) holds, both effective thermal conductivities are equal. For the HS scheme this is not true. The first-order bounds for the effective thermal conductivity (cf. Appendix B.2) are also included as limit cases

$$\lim_{\lambda_0 \to 0} \bar{\boldsymbol{\lambda}}^{q} = \lim_{\lambda_0 \to 0} \bar{\boldsymbol{\lambda}}^{W} = \left\langle \lambda^{-1} \right\rangle^{-1} \boldsymbol{I} \tag{5.40}$$

$$\lim_{\lambda_0 \to \infty} \bar{\boldsymbol{\lambda}}^{q} = \lim_{\lambda_0 \to \infty} \bar{\boldsymbol{\lambda}}^{W} = \langle \lambda \rangle \, \boldsymbol{I}. \tag{5.41}$$

The HS upper or lower bounds are obtained if the thermal conductivity of the comparison material is chosen the way that the expression $\lambda_\alpha - \lambda_0$ is negative or positive for all phases, respectively. Thus, an infinite amount of bounds can be found. The optimal upper and lower bounds result from the minimum λ_0 that the expression $\lambda_\alpha - \lambda_0$ is negative and from the maximum λ_0 that the expression $\lambda_\alpha - \lambda_0$ is positive, respectively.

A self-consistent (SC) estimate for the effective thermal conductivity can be derived choosing the thermal conductivity of the comparison material to be equal to the effective thermal conductivity. Therefore, nonlinear equations for both methods for the determination of the effective thermal conductivity are obtained

$$\bar{\lambda}^q = \left\langle \frac{\lambda}{\lambda + 2\bar{\lambda}^q} \right\rangle \left\langle \frac{1}{\lambda + 2\bar{\lambda}^q} \right\rangle^{-1}, \tag{5.42}$$

$$\bar{\lambda}^W = \left\langle \frac{\lambda}{\left(\lambda + 2\bar{\lambda}^W\right)^2} \right\rangle \left\langle \frac{1}{\lambda + 2\bar{\lambda}^W} \right\rangle^{-2}. \tag{5.43}$$

On the left-hand side of Figure 5.1, the first-order bounds, the optimal HS bounds, the limits of the HS effective thermal conductivity and the SC estimate for an austenitic-martensitic microstructure of the steel 22MnB5 are shown. On the right-hand side, the two methods for the determination of the HS bounds and the SC estimate are depicted.

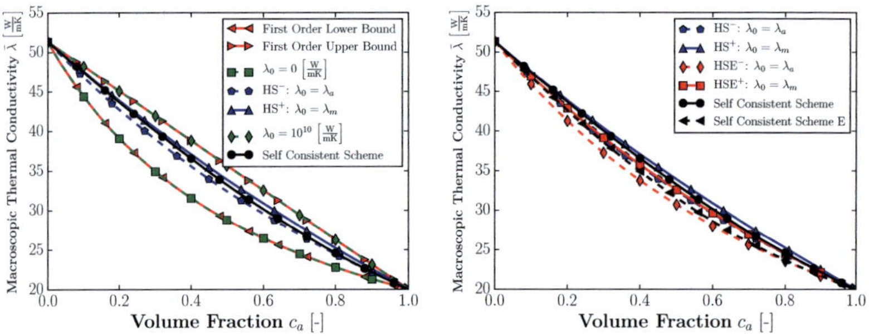

Figure 5.1: Bounds and estimates for the effective thermal conductivity for an austenitic-martensitic microstructure of the steel 22MnB5 (left) and comparison of the determination methods of the effective thermal conductivity (E marks the method from the energy consideration) (right)

The effective thermal conductivity determined from the energy consideration seems to span a regime for the macroscopic thermal conductivity equivalent to the regime of the mechanically derived equations, but it is set off towards the bottom.

The thermal HS localization relation is derived by introducing the heat flux polarization tensor

$$\boldsymbol{\pi}_{\alpha} = \boldsymbol{q}_{\alpha} - \boldsymbol{q}_0 = \delta\boldsymbol{\lambda}_{\alpha}\boldsymbol{g}_{\alpha}, \tag{5.44}$$

i.e. the difference between the heat flux in the phase α and the heat flux in the homogeneous comparison material. Thereby, the abbreviation $\delta\boldsymbol{\lambda} = \boldsymbol{\lambda}_{\alpha} - \boldsymbol{\lambda}_0$ has been introduced. With the stress polarization, the thermal HS functional reads (Hashin and Shtrikman, 1962b)

$$\mathcal{F}^{\theta} = \left\langle \frac{1}{2}\boldsymbol{\pi} \cdot \left(\delta\boldsymbol{\lambda}^{-1} + \boldsymbol{P}_0\right)\boldsymbol{\pi} - \frac{1}{2}\boldsymbol{\pi} \cdot \boldsymbol{P}_0 \left\langle \boldsymbol{\pi} \right\rangle - \boldsymbol{\pi} \cdot \bar{\boldsymbol{g}} \right\rangle. \tag{5.45}$$

With the assumptions stated above, the determination of the stationary point via

$$\delta\mathcal{F}^{\theta} = \frac{\partial F^{\theta}}{\partial \boldsymbol{\pi}_{\alpha}} \cdot \delta\boldsymbol{\pi}_{\alpha} = 0 \tag{5.46}$$

leads to the thermal phase interaction law

$$\boldsymbol{q}_{\alpha} - \bar{\boldsymbol{q}} = \boldsymbol{L}^{\theta}\left(\boldsymbol{g}_{\alpha} - \bar{\boldsymbol{g}}\right), \tag{5.47}$$

where the thermal Hill's constraint tensor is defined by $\boldsymbol{L}^{\theta} = \boldsymbol{\lambda}_0 - \boldsymbol{P}_0^{-1}$. Using Fourier's law for the phases and assuming an isotropic comparison material, the temperature gradient localization law

$$\boldsymbol{g}_{\alpha} = \boldsymbol{A}_{\alpha}\bar{\boldsymbol{g}} = \boldsymbol{R}_{\alpha}\left\langle \boldsymbol{R} \right\rangle^{-1}\bar{\boldsymbol{g}}, \tag{5.48}$$

$$\boldsymbol{R}_{\alpha} = \left(\boldsymbol{I} + \boldsymbol{P}_0\delta\boldsymbol{\lambda}_{\alpha}\right)^{-1}, \tag{5.49}$$

$$\boldsymbol{P}_0 = \frac{1}{3\lambda_0}\boldsymbol{I} \tag{5.50}$$

for each phase is obtained. Furthermore, by averaging Fourier's law for the phases, using the thermal Hill condition, and comparison with the macroscopic Fourier's law, the effective HS thermal conductivity

$$\bar{\boldsymbol{\lambda}}^q = \langle \boldsymbol{A}\boldsymbol{\lambda} \rangle = \langle \boldsymbol{\lambda}\boldsymbol{A} \rangle \tag{5.51}$$

can be derived. Considering the stored thermal energies, the other form of the macroscopic thermal conductivity

$$\bar{\boldsymbol{\lambda}}^W = \left\langle \boldsymbol{A}^{\mathrm{T}}\boldsymbol{\lambda}\boldsymbol{A} \right\rangle \tag{5.52}$$

is obtained, which is equivalent to equation (5.51) if and only if the Hill condition is valid.

5.3.2 The Hashin-Shtrikman-Type Estimate of Mechanical Properties[2]

For the estimation of the macroscopic thermomechanical behavior, the eigenstrain concept is followed, where the inelastic and thermal strains are considered as eigenstrains (c.f. Neumann and Böhlke, 2016). Following the bounding technique by Willis (1977), a homogeneous comparison material with the stiffness $\mathbb{C}_0$ is introduced and the stress polarization

$$\boldsymbol{p}(\boldsymbol{x}) = \boldsymbol{\sigma}(\boldsymbol{x}) - \mathbb{C}_0[\boldsymbol{\varepsilon}(\boldsymbol{x})], \tag{5.53}$$

is defined. Neglecting the body force on micro scale, the mechanical boundary value problem (BVP) on the micro scale is given in terms of the stress polarization by

$$\operatorname{div}\left(\mathbb{C}_0[\boldsymbol{\varepsilon}(\boldsymbol{x})]\right) + \operatorname{div}\left(\boldsymbol{p}(\boldsymbol{x})\right) = \boldsymbol{0}, \tag{5.54}$$

$$\bar{\boldsymbol{\varepsilon}} = \langle \boldsymbol{\varepsilon} \rangle . \tag{5.55}$$

[2] The Hashin-Shtrikman approach, which is described in this chapter, is based on the paper "Hashine-Shtrikman type mean field model for the two-scale simulation of the thermomechanical processing of steel" (Neumann and Böhlke, 2016) and a reduced formulation for isotropic phase behavior is introduced.

The second term of equation (5.54) can be interpreted as a body force on the micro scale. The solution of this BVP is formally given by

$$\varepsilon(\boldsymbol{x}) = \bar{\varepsilon} - \int_{V'} \mathbb{G}(\boldsymbol{x}' - \boldsymbol{x})[\boldsymbol{p}(\boldsymbol{x}') - \langle \boldsymbol{p} \rangle] \mathrm{V}', \qquad (5.56)$$

where the fourth-order tensor $\mathbb{G}$ includes the second derivatives of infinite body's Green's function (see, e.g., Ponte Castañeda and Suquet, 1998). The assumptions of a/an

(i) Constant volume fractions of the phases,

(ii) Piece-wise constant stress polarization,

(iii) Statistical homogeneity and no long-range order,

(iv) Isotropic comparison material, and

(v) Isotropic two-point statistics,

imply the following relation for the phase average of the strain (see, e.g., Jöchen and Böhlke, 2012; Willis, 1977)

$$\varepsilon_\alpha = \bar{\varepsilon} + \frac{1}{c_\beta} \sum_{\beta=1}^{N} \mathbb{G}_{\alpha\beta}[\boldsymbol{p}_\beta], \qquad (5.57)$$

where the fourth-order tensor $\mathbb{G}_{\alpha\beta} = c_\alpha(\delta_\beta - c_\beta)\mathbb{P}_0$ is given by Willis (1977). Thereby, δ_β denotes the Kronecker-Delta and $\mathbb{P}_0$ the fourth-order polarization tensor. For an isotropic comparison material with the stiffness

$$\mathbb{C}_0 = 3K_0 \mathbb{P}^{1,\,\mathrm{iso}} + 2G_0 \mathbb{P}^{2,\,\mathrm{iso}}, \qquad (5.58)$$

the polarization tensor reads (see, e.g., Dederichs and Zeller, 1973)

$$\mathbb{P}_0 = \frac{1}{3K_0 + 4G_0} \mathbb{P}^{1,\,\mathrm{iso}} + \frac{2}{10G_0} \frac{3K_0 + 6G_0}{3K_0 + 4G_0} \mathbb{P}^{2,\,\mathrm{iso}}. \qquad (5.59)$$

With this polarization tensor, a phase interaction law for the determination of the phase stresses or strains is obtained from equation (5.57)

$$\boldsymbol{\sigma}_\alpha - \langle \boldsymbol{\sigma} \rangle = \mathbb{L}[\boldsymbol{\varepsilon}_\alpha - \langle \boldsymbol{\varepsilon} \rangle], \tag{5.60}$$

which is a set of N nonlinear equations. Thereby, $\mathbb{L} = \mathbb{C}_0 - \mathbb{P}_0^{-1}$ denotes the Hill's constraint tensor. Note that in the two limits

$$\|\mathbb{L}\| \to \begin{cases} 0 & \text{if} \quad \|\mathbb{C}_0\| \to 0 \\ \infty & \text{if} \quad \|\mathbb{C}_0\| \to \infty \end{cases}, \tag{5.61}$$

the simple first-order bounds of Sachs and Taylor are obtained (Böhlke et al., 2014; Ponte Castañeda and Suquet, 1998).

Introducing Hooke's law (5.18) into the phase interaction law (5.60), one obtains a strain localization relation for each phase of the form

$$\boldsymbol{\varepsilon}_\alpha = \mathbb{A}_\alpha \left[\bar{\boldsymbol{\varepsilon}}\right] - \boldsymbol{A}_\alpha^E, \tag{5.62}$$

where $\mathbb{A}_\alpha$ is referred to as the fourth-order localization tensor and $\boldsymbol{A}_\alpha$ as the eigenstrain influence tensor of phase α. The fourth-order localization tensor and the nonlinear contribution due to eigenstrains $\boldsymbol{A}_\alpha^E$ are given by

$$\mathbb{A}_\alpha = \mathbb{R}_\alpha \langle \mathbb{R} \rangle^{-1}, \qquad \boldsymbol{A}_\alpha^E = -\mathbb{R}_\alpha \mathbb{C}_\alpha \left[\boldsymbol{\varepsilon}_\alpha^E\right] + \mathbb{A}_\alpha \left\langle \mathbb{R}\mathbb{C}\left[\boldsymbol{\varepsilon}^E\right]\right\rangle, \tag{5.63}$$

where $\mathbb{R}_\alpha = \left(\mathbb{P}_0^{-1} + \delta\mathbb{C}_\alpha\right)^{-1}$ and $\delta\mathbb{C}_\alpha = \mathbb{C}_\alpha - \mathbb{C}_0$ hold (Jöchen and Böhlke, 2012; Jöchen, 2013). By averaging, one can directly see, that the normalization conditions

$$\langle \mathbb{A} \rangle = \mathbb{I}, \qquad \left\langle \boldsymbol{A}^E \right\rangle = \mathbf{0} \tag{5.64}$$

are fulfilled.

Since the eigenstrain ε_{α}^{E}, in particular the plastic strain and TRIP strain, are dependent on the stress states of all phases, the localization relation (5.62) is an implicit set of nonlinear equations of the phase strains. Therefore, for the solution of the localization relation (5.62), an iterative solution scheme has to be used.

Note, that the phase interaction law (5.60) is the simplified relation, due to assumption (i)-(v), of equation (5.57). The difference between equation (5.57) and the strain localization relation (5.62) results from the introduction of the material law.

"Note, that for each choice of the stiffness of the comparison material $\mathbb{C}_0$, one obtains admissible estimates for the macroscopic material behavior in-between the first-order bounds. The physical interpretation of the estimate is, that, for a magnitude of the stiffness of the comparison material towards infinite, the strain fluctuations on the micro scale vanish. On the other hand, for a magnitude of stiffness of the comparison material of zero, the stress fluctuations are suppressed. Therefore, the stiffness of the comparison material is a parameter to influence the fluctuations of the stress and strain fields, and is set to get the best approximation of the macroscopic material behavior" (c.f. Neumann and Böhlke, 2016).

Note, that assumption (i) demands constant volume fractions during the entire process. However, the volume fractions vary during the process due to both volumetric strains and phase transformation effects discussed in Section 4.4. In a microstructure, which consists of different phases with differing bulk moduli or thermal expansion coefficients, the volume increases of each phase are different, leading to varying volume fractions. This effect is discussed in Appendix C.

The macroscopic stress and strain are determined by direct averaging

$$\bar{\sigma} = \langle \sigma \rangle , \qquad \bar{\varepsilon} = \langle \varepsilon \rangle , \tag{5.65}$$

respectively. Using the Hill-Mandel condition (Hill, 1963)

$$\langle \boldsymbol{\sigma} \cdot \boldsymbol{\varepsilon} \rangle = \langle \boldsymbol{\sigma} \rangle \cdot \langle \boldsymbol{\varepsilon} \rangle \,, \tag{5.66}$$

which is valid for homogeneous Neumann boundary conditions, linear Dirichlet boundary conditions, periodic boundary conditions, and an infinitely extended volume, the macroscopic stiffness tensor (see, e.g., Jöchen, 2013)

$$\bar{\mathbb{C}} = \langle \mathbb{C}\mathbb{A} \rangle \tag{5.67}$$

can be derived. The macroscopic eigenstrain tensor is obtained by the Mandel-Levin equation (Levin, 1967; Mandel, 1965) in terms of the strain localization tensor

$$\bar{\boldsymbol{\varepsilon}}^E = \bar{\mathbb{C}}^{-1} \left\langle \mathbb{A}^{\mathrm{T_H}} \mathbb{C} \left[\boldsymbol{\varepsilon}^E \right] \right\rangle . \tag{5.68}$$

Reduction for Elastic-Isotropic Phases

In the special case of elastic-isotropic phases (5.19) on the micro structure, the localization tensors of the HS-type estimate (5.62) can be determined by $2N$ parameters via

$$\mathbb{A}_\alpha = \frac{\tau_\alpha}{\langle \tau \rangle} \mathbb{P}^{1,\,\mathrm{iso}} + \frac{\varsigma_\alpha}{\langle \varsigma \rangle} \mathbb{P}^{2,\,\mathrm{iso}} , \tag{5.69}$$

where N is the number of phases, τ_α the spheric, and ς_α the deviatoric parameter of the localization tensor. The localization parameters are given by

$$\tau_\alpha = \frac{3K_\alpha}{3K_0 + 4G_0 + 3\delta K_\alpha}, \qquad \varsigma_\alpha = \frac{2G_\alpha}{\frac{5G_0(3K_0+6G_0)}{3K_0+4G_0} + 2\delta G_\alpha}, \tag{5.70}$$

where $\delta K_\alpha = K_\alpha - K_0$ and $\delta G_\alpha = G_\alpha - G_0$ have been defined. The parameters' range of values is $[0, 1]$ depending on the stiffness of the comparison material. For a vanishing stiffness of the comparison material, the localization parameters are zero and the first-order bound of

Sachs is obtained. For a stiffness of the comparison material towards infinity, the localization parameters are one and the first-order bound of Taylor is obtained.

Note, that the $2N$ parameters are not arbitrary, but they are dependent on the two elasticity parameters of the stiffness tensor of the comparison material.

With the reduced strain localization tensor, the localization relation (5.62) can be decomposed into two separate equations, an equation for the spherical part

$$\varepsilon_\alpha^\circ = \frac{\tau_\alpha}{K_\alpha}\left\langle\frac{\tau}{K}\right\rangle^{-1}\left(\bar{\varepsilon}^\circ + K_\alpha\left\langle\frac{\tau}{K}\right\rangle\varepsilon_\alpha^{E\circ} - \left\langle\tau\varepsilon^{E\circ}\right\rangle\right) \tag{5.71}$$

and an equation for the deviatoric part of the phase strain tensor

$$\varepsilon_\alpha' = \frac{\varsigma_\alpha}{G_\alpha}\left\langle\frac{\varsigma}{G}\right\rangle^{-1}\left(\bar{\varepsilon}' + G_\alpha\left\langle\frac{\varsigma}{G}\right\rangle\varepsilon_\alpha^{E\,\prime} - \left\langle\varsigma\,\varepsilon^{E\,\prime}\right\rangle\right). \tag{5.72}$$

For a microstructure consisting of phases with equal isotropic elasticity, e.g., ferritic polycrystal with anisotropic plasticity, where the phases are distinguished by the initial orientation of the grains (discussed in, e.g., Jöchen, 2013), a further simplification of the localization relation is possible. In this case, the localization relation is described by two parameters. The spherical part and the deviatoric part of the localization relation read

$$\varepsilon_\alpha^\circ = \bar{\varepsilon}^\circ + \tau\left(\varepsilon_\alpha^{E\circ} - \left\langle\varepsilon^{E\circ}\right\rangle\right) \tag{5.73}$$

and

$$\varepsilon_\alpha' = \bar{\varepsilon}' + \varsigma\left(\varepsilon_\alpha^{E\,\prime} - \left\langle\varepsilon^{E\,\prime}\right\rangle\right), \tag{5.74}$$

respectively. Since the range of values of both parameters depending on the stiffness of the comparison material is $[0, 1]$, the two parameters control the influence of the eigenstrain fluctuation in the phases on the phase strain.

Note, that, in this case, both parameters can be chosen arbitrarily. The elasticity parameters of stiffness tensor of the comparison material can be determined by solving the set of equations

$$\tau = \frac{3K}{3K_0 + 4G_0 + 3\delta K}, \qquad \varsigma = \frac{2G}{\frac{5G_0(3K_0+6G_0)}{3K_0+4G_0} + 2\delta G}. \tag{5.75}$$

Extension to a Self-Consistent Scheme

A SC scheme for a polycrystal without distinct phase can be directly derived from the HS-type estimation given above, by choosing the stiffness of the comparison material equal to the macroscopic stiffness. Since the localization tensors are a function of the stiffness of the comparison material, a nonlinear equation for the determination of macroscopic stiffness tensor is obtained from relation (5.67)

$$\bar{\mathbb{C}} = \left\langle \mathbb{C} \left(\mathbb{P}_0(\bar{K},\bar{G})^{-1} + \mathbb{C} - \bar{\mathbb{C}}\right)^{-1} \right\rangle \left\langle \mathbb{P}_0(\bar{K},\bar{G})^{-1} + \mathbb{C} - \bar{\mathbb{C}} \right\rangle^{-1}. \tag{5.76}$$

This means, that, in addition to the nonlinear localization relation (5.62), a further nonlinear relation has to be solved to be able to find the phase strains.

For the special case of elastic-isotropic phases, the SC estimate can also be reduced to a nonlinear system of two coupled scalar equations

$$\bar{K} = \frac{\langle K\tau \rangle}{\langle \tau \rangle} = \frac{\left\langle \frac{K^2}{3\bar{K}+4\bar{G}+3\left(K-\bar{K}\right)} \right\rangle}{\left\langle \frac{K}{3\bar{K}+4\bar{G}+3\left(K-\bar{K}\right)} \right\rangle}, \quad \bar{G} = \frac{\langle G\varsigma \rangle}{\langle \varsigma \rangle} = \frac{\left\langle \frac{G^2}{\frac{5\bar{G}(3\bar{K}+6\bar{G})}{3\bar{K}+4\bar{G}}+2(G-\bar{G})} \right\rangle}{\left\langle \frac{G}{\frac{5\bar{G}(3\bar{K}+6\bar{G})}{3\bar{K}+4\bar{G}}+2(G-\bar{G})} \right\rangle}. \tag{5.77}$$

For a microstructure consisting of phases with equal isotropic elasticity, the effective stiffness corresponds to the isotropic phases' stiffness.

5.3.3 The Hill-Mandel Lemma-Type Relations[3]

The balance of linear momentum (3.32) holds on regular points. Therefore, for a phase-wise constant material behavior, this balance reads

$$\operatorname{div}(\boldsymbol{\sigma}_\alpha) = \operatorname{div}\left(\mathbb{C}_\alpha\left[\boldsymbol{\varepsilon}_\alpha - \boldsymbol{\varepsilon}_\alpha^E\right]\right) = \operatorname{div}\left(\mathbb{C}_\alpha\mathbb{A}_\alpha\left[\bar{\boldsymbol{\varepsilon}}\right] - \mathbb{C}_\alpha\left[\boldsymbol{A}_\alpha^E + \boldsymbol{\varepsilon}_\alpha^E\right]\right) = \mathbf{0}, \tag{5.78}$$

where Hooke's law and the strain localization realization (5.62) have been introduced. The latter expression shows, that for an elastic material behavior and, thus, for any possible thermomechanical load, the expression $\mathbb{C}_\alpha\mathbb{A}_\alpha\left[\bar{\boldsymbol{\varepsilon}}\right]$ is divergence-free. Due to the linearity of the divergence operator, the expression $\mathbb{C}_\alpha\left[\boldsymbol{A}_\alpha^E + \boldsymbol{\varepsilon}_\alpha^E\right]$ also has to be divergence-free for any thermomechanical load.

Since the integrability property is linear, i.e. the sum of two integrable quantities and a linear mapping of integrable quantities are integrable itself, the expression $\mathbb{A}_\alpha\left[\bar{\boldsymbol{\varepsilon}}\right]$ is integrable. Thus, since the phase strain $\boldsymbol{\varepsilon}_\alpha$ is integrable, from the localization relation (5.62)

$$\boldsymbol{A}_\alpha^E = \boldsymbol{\varepsilon}_\alpha - \mathbb{A}_\alpha\left[\bar{\boldsymbol{\varepsilon}}\right], \tag{5.79}$$

it follows that the nonlinear contribution to the localization relation $\boldsymbol{A}_\alpha^E$ is integrable for any possible loading. Due to Mandel-Hill's lemma, i.e. the average of a scalar product of a divergence-free and an integrable quantity is equal to the scalar product of the averages of these quantities, the two relations hold

$$\left\langle\mathbb{C}\left[\boldsymbol{A}^E + \boldsymbol{\varepsilon}^E\right]\cdot\boldsymbol{A}^E\right\rangle = \left\langle\mathbb{C}\left[\boldsymbol{A}^E + \boldsymbol{\varepsilon}^E\right]\right\rangle\cdot\left\langle\boldsymbol{A}^E\right\rangle = \mathbf{0}, \tag{5.80}$$

$$\left\langle\mathbb{C}\mathbb{A}\left[\bar{\boldsymbol{\varepsilon}}\right]\cdot\boldsymbol{A}^E\right\rangle = \left\langle\mathbb{C}\mathbb{A}\left[\bar{\boldsymbol{\varepsilon}}\right]\right\rangle\cdot\left\langle\boldsymbol{A}^E\right\rangle = \mathbf{0}. \tag{5.81}$$

[3] The Hill-Mandel lemma-type relations are discussed in the paper "Hashine-Shtrikman type mean field model for the two-scale simulation of the thermomechanical processing of steel" (Neumann and Böhlke, 2016). For the sake of completeness the discussion is represented.

5.4 Special Form of the Heat Equation[4]

The macroscopic Helmholtz free energy is obtained by the ensemble averaged from the Helmholtz free energy

$$\bar{\psi} = \langle \psi \rangle = \left\langle \psi^{te} \right\rangle + \left\langle \psi^{i} \right\rangle . \tag{5.82}$$

The thermoelastic part of the Helmholtz free energy ψ^{te}_{α} is defined by (cf. equation (5.17))

$$\psi^{te}_{\alpha} = \frac{1}{2\rho_{\alpha}} \left(\varepsilon_{\alpha} - \varepsilon^{E}_{\alpha} \right) \cdot \mathbb{C}_{\alpha} \left[\varepsilon_{\alpha} - \varepsilon^{E}_{\alpha} \right] - \frac{9K_{\alpha}}{2\rho_{\alpha}} \alpha^{2}_{\alpha} \Delta\theta^{2} - \kappa^{c}_{\alpha} \left(\theta \ln \left(\frac{\theta}{\theta_{0}} \right) - \Delta\theta \right), \tag{5.83}$$

where, for an isotropic elastic material behavior and an isotropic temperature expansion, the middle term has been evaluated. With the Hill-Mandel lemma-type relations (5.80) and (5.81), the localization relation (5.62), and the Mandel-Levin formula (5.68), as well as the thermoelastic part of the macroscopic Helmholtz free energy reads

$$\bar{\psi}^{te} = \frac{1}{2\bar{\rho}} \left(\bar{\varepsilon} - \bar{\varepsilon}^{E} \right) \cdot \bar{\mathbb{C}} \left[\bar{\varepsilon} - \bar{\varepsilon}^{E} \right] - \frac{1}{2\bar{\rho}} \bar{\varepsilon}^{E} \cdot \bar{\mathbb{C}} \left[\bar{\varepsilon}^{E} \right] - \frac{9}{2\bar{\rho}} \left\langle K \alpha^{2} \right\rangle \Delta\theta^{2} - \frac{1}{2\bar{\rho}} \left\langle \varepsilon^{E} \cdot \mathbb{C} \left[\boldsymbol{A}^{E} + \varepsilon^{E} \right] \right\rangle - \left\langle \kappa^{c} \right\rangle \left(\theta \ln \left(\frac{\theta}{\theta_{0}} \right) - \Delta\theta \right). \tag{5.84}$$

Thereby, similar to the eigenstrain tensor on the micro scale (5.21), the macroscopic eigenstrain tensor occurs as the sum of the macroscopic plastic strain tensor $\bar{\varepsilon}^{p}$, the macroscopic TRIP strain tensor $\bar{\varepsilon}^{tr}$, the macroscopic transformation strain tensor $\bar{\varepsilon}^{t}$, and the macroscopic thermal

[4] The discussion of the special form of the heat equation is based on the paper "Hashine-Shtrikman type mean field model for the two-scale simulation of the thermomechanical processing of steel" (Neumann and Böhlke, 2016). For the sake of completeness the discussion is represented.

expansion

$$\bar{\varepsilon}^E = \bar{\varepsilon}^p + \bar{\varepsilon}^t + \bar{\alpha}\Delta\theta\boldsymbol{I}. \tag{5.85}$$

Using the Mandel-Levin formula (5.68), and the reduced localization relation for isotropic elastic phases (5.69), the macroscopic thermal expansion coefficient reads

$$\bar{\alpha} = \frac{\langle \alpha K \tau \rangle}{\langle K \tau \rangle}. \tag{5.86}$$

In an equivalent way, the macroscopic plastic strain, the macroscopic TRIP strain, and the macroscopic mean transformation volume change

$$\bar{\varepsilon}^p = \frac{1}{\langle G\varsigma \rangle}\langle G\varsigma\varepsilon^p \rangle, \qquad \bar{\varepsilon}^{tr} = \frac{1}{\langle G\varsigma \rangle}\langle G\varsigma\varepsilon^{tr} \rangle, \qquad \bar{\zeta} = \frac{\langle \zeta\mathcal{H}(c) K \tau \rangle}{\langle K \tau \rangle} \tag{5.87}$$

can be determined.

The macroscopic inelastic part of the Helmholtz free energy is determined by averaging of equation (5.23) for the Voce-type hardening and (5.24) for the Swift-type hardening. Altogether, the macroscopic Helmholtz free energy occurs as a function of the macroscopic elastic strain defined by

$$\bar{\varepsilon}^e = \bar{\varepsilon} - \bar{\varepsilon}^p - \bar{\varepsilon}^{tr} - \bar{\varepsilon}^t, \tag{5.88}$$

the eigenstrains, accumulated plastic strains, and volume fractions of all phases, and the temperature

$$\bar{\psi} = \bar{\psi}\left(\bar{\varepsilon}^e, \varepsilon_1^E, \ldots, \varepsilon_N^E, \varepsilon_1^p, \ldots, \varepsilon_N^p, c_1, \ldots, c_N, \theta\right). \tag{5.89}$$

Thus, the total differential of the macroscopic Helmholtz free energy reads

$$\mathrm{d}\bar{\psi} = \frac{\partial\bar{\psi}}{\partial\bar{\varepsilon}^e}\,\mathrm{d}\bar{\varepsilon}^e + \sum_{\alpha\in\mathcal{PH}}\left(\frac{\partial\bar{\psi}}{\partial\varepsilon_\alpha^E}\,\mathrm{d}\varepsilon_\alpha^E + \frac{\partial\bar{\psi}}{\partial\varepsilon_\alpha^p}\,\mathrm{d}\varepsilon_\alpha^p + \frac{\partial\bar{\psi}}{\partial c_\alpha}\,\mathrm{d}c_\alpha\right) + \frac{\partial\bar{\psi}}{\partial\theta}\,\mathrm{d}\theta. \tag{5.90}$$

The term containing the partial derivation with respect to the volume fractions of the phases can be interpreted by the latent heats released or consumed during phase transformation. In this work, the case of the decomposition of one phase into $N-1$ phases is referred to as an one phase decomposition. Without loss of generality, the decomposed phase is referred to as phase 1. Using $\mathrm{d}c_1 = -\sum_{\alpha=2}^{N} \mathrm{d}c_\alpha$, the volume fraction related term reads

$$\begin{aligned}\sum_{\alpha\in\mathcal{PH}} \frac{\partial\bar{\psi}}{\partial c_\alpha}\,\mathrm{d}c_\alpha = \sum_{\alpha\in\mathcal{PH}} \psi_\alpha\,\mathrm{d}c_\alpha &= \sum_{\alpha\in\mathcal{PH}\backslash 1} (\psi_\alpha - \psi_1)\,\mathrm{d}c_\alpha \\ &= -\frac{1}{\bar{\rho}} \sum_{\alpha\in\mathcal{PH}\backslash 1} L_\alpha\,\mathrm{d}c_\alpha,\end{aligned} \tag{5.91}$$

where the dissipated Helmholtz free energy difference between the parent phase and the growing phase $L_\alpha = \bar{\rho}(\psi_1 - \psi_\alpha)$ is referred to as latent heat. The remaining partial derivations of the total differential of the Helmholtz free energy (5.90) are shown in Appendix D.

Considering the total differential of the macroscopic Helmholtz free energy (5.90), the heat equation on the macro scale in terms of the macroscopic Helmholtz free energy reads

$$\bar{\rho}\dot{\bar{\psi}} - \mathrm{div}\,(\bar{\boldsymbol{q}}) = \bar{\boldsymbol{\sigma}}\cdot\dot{\bar{\varepsilon}} - \bar{\rho}\dot{\theta}\bar{\eta} - \bar{\rho}\theta\dot{\bar{\eta}} + \bar{w}. \tag{5.92}$$

Fourier's law on macro scale is given $\bar{\boldsymbol{q}} = -\bar{\boldsymbol{\lambda}}\bar{\boldsymbol{g}}$, where the effective thermal conductivity is discussed in Section 5.3.1.

With the potential relations for the macroscopic stress and macroscopic entropy

$$\bar{\boldsymbol{\sigma}} = \bar{\rho}\frac{\partial\bar{\psi}}{\partial\bar{\varepsilon}^e}, \qquad \bar{\eta} = -\frac{\partial\bar{\psi}}{\partial\theta} \tag{5.93}$$

derivable from the Clausius-Duhem inequality on the macro scale, the heat equation on the macro scale reads

$$-\bar{\rho}\bar{\kappa}\dot{\theta} - \operatorname{div}\left(\bar{\lambda}\operatorname{grad}(\theta)\right) = \sum_{\alpha\in\mathcal{PH}\setminus 1} L_\alpha \dot{c}_\alpha + \bar{\boldsymbol{\sigma}} \cdot \dot{\bar{\varepsilon}}^i + \bar{w} -$$
$$\bar{\rho} \sum_{\alpha\in\mathcal{PH}} \left(\frac{\partial\bar{\psi}^{te}}{\partial\varepsilon_\alpha^E} \cdot \dot{\varepsilon}_\alpha^E + \frac{\partial\bar{\psi}^i}{\partial\varepsilon_\alpha^p} \dot{\varepsilon}_\alpha^p \right) - \bar{\rho}\theta \frac{\mathrm{d}}{\mathrm{d}t} \sum_{\alpha\in\mathcal{PH}} \left(\frac{\partial\bar{\psi}^{te}}{\partial\varepsilon_\alpha^E} \cdot \bar{\boldsymbol{\alpha}} + \frac{\partial\bar{\psi}}{\partial c_\alpha} \frac{\mathrm{d}c_\alpha}{\mathrm{d}\theta} \right), \tag{5.94}$$

where the macroscopic specific heat capacity and the macroscopic inelastic strain tensor are defined by

$$\bar{\kappa} = \theta\partial^2\bar{\psi}/\partial\theta^2 \tag{5.95}$$

and

$$\bar{\varepsilon}^i = \bar{\varepsilon}^p + \bar{\varepsilon}^{tr} + \bar{\varepsilon}^t, \tag{5.96}$$

respectively.

Note, the diffusion-driven phase transformation laws are explicitly dependet on the temperature. As it can be seen in the JMAK equation (4.75), both JMAK parameters a_α and b_α are dependent on the temperature. The KV equation is directly dependent on the temperature via the temperature influence function (4.81). Due to the dependence of the Heaviside function on θ, the same is true for the diffusionless transformation.

Simplification of the Heat Equation in the Special Case of Hot Stamping

The developed model is used to consider thermomechanical problems with high heat transport through the boundaries, e.g., quenching of austenitized parts. In such a process, the influence of the heat transfer is the significant quantity in the development of the sheet metal temperature. Thus, as a first reasonable approach, the contributions to the temperature development due to mechanical dissipation are neglected leading to

$$\bar{\rho}\bar{\kappa}\dot{\theta} - \operatorname{div}\left(\bar{\lambda}\operatorname{grad}(\theta)\right) = \sum_{\alpha \in \mathcal{PH}\setminus 1} L_\alpha \dot{c}_\alpha. \tag{5.97}$$

In the case of evolving phase fractions, the macroscopic specific heat capacity reads

$$\bar{\kappa} = -\theta \frac{\partial^2 \bar{\psi}}{\partial \theta^2} = \theta \frac{\partial \bar{\eta}}{\partial \theta} = \theta \sum_{\alpha \in \mathcal{PH}} \frac{\partial c_\alpha}{\partial \theta} \eta_\alpha + \langle \kappa^c \rangle, \tag{5.98}$$

where the phases' specific heat capacity is defined by $\kappa_\alpha = -\theta \partial^2 \psi_\alpha / \partial \theta^2$. For the sake of simplicity, the macroscopic heat capacity is set to be only the volume average $\bar{\kappa} = \langle \kappa^c \rangle$.

5.5 Aspects of the Thermodynamical Consistency on the Macroscale[5]

To be consistent, the model has to fulfill the CD inequality on the macro scale. The derivation of the constitutive behavior of the phases, taking the Clausius-Duhem inequality on micro scale into account, defines a thermodynamically consistent phase behavior. The thermodynamic consistency of the overall macroscopic material behavior is verified by comparing the volume average of the dissipation on the micro scale with the macroscopic dissipation during the thermomechanical loading process. Considering only the mechanical dissipation, the volume average of the dissipation equation (3.34) reads

$$\langle \tilde{D} \rangle = -\bar{\rho} \langle \dot{\psi} \rangle - \bar{\rho} \langle \eta \rangle \dot{\theta} + \langle \boldsymbol{\sigma} \rangle \cdot \langle \dot{\boldsymbol{\varepsilon}} \rangle \geq 0, \tag{5.99}$$

[5] The thermodynamic consistency of the overall model is discussed in the paper "Hashine-Shtrikman type mean field model for the two-scale simulation of the thermomechanical processing of steel" (Neumann and Böhlke, 2016). For the sake of completeness the discussion is represented.

where the Hill-Mandel relation (5.66) has been used. The corresponding macroscopic dissipation is given by

$$\bar{\tilde{D}} = -\bar{\rho}\dot{\bar{\psi}} - \bar{\rho}\bar{\eta}\dot{\theta} + \bar{\boldsymbol{\sigma}} \cdot \dot{\bar{\boldsymbol{\varepsilon}}} \tag{5.100}$$

which leads, by use of equation (5.99), to

$$\bar{\tilde{D}} = \langle \tilde{D} \rangle - \bar{\rho} \sum_{\alpha \in \mathcal{PH}} \left(\dot{c}_\alpha + \frac{\partial c_\alpha}{\partial \theta} \right) \psi_\alpha + \bar{\boldsymbol{\sigma}} \cdot \sum_{\alpha \in \mathcal{PH}} \dot{c}_\alpha \boldsymbol{\varepsilon}_\alpha . \tag{5.101}$$

Thereby, the relation for averaged quantities in the case of evolving volume fractions (3.41) has been used. Equation (5.101) shows that, in the general case of evolving microstructures, the dissipation is not only the volume average of the dissipation on the microstructure. It also contains contributions due to Helmholtz free energy changes and changes of the mechanical dissipation if volume changes take place on the micro scale.

Note, that in the special case of constant volume fractions during the process, since the second and the third term in equation (5.101) vanish and the phases behave thermodynamically consistently, the overall thermodynamic consistency is directly proven.

In the case of a one phase decomposition, the second term in equation (5.101) can be rearranged into

$$-\bar{\rho} \sum_{\alpha \in \mathcal{PH}} \left(\dot{c}_\alpha + \frac{\partial c_\alpha}{\partial \theta} \right) \psi_\alpha = \sum_{\alpha \in \mathcal{PH} \backslash 1} \left(\dot{c}_\alpha + \frac{\partial c_\alpha}{\partial \theta} \right) L_\alpha , \tag{5.102}$$

where the definition of the latent heat (5.91) has been used. The released heat during such a transformation as well as the change of the volume fractions c_α ($\alpha \in \{2, \dots N\}$) are positive, so that this term fulfills the dissipation inequality

$$\sum_{\alpha \in \mathcal{PH} \backslash 1} \left(\dot{c}_\alpha + \frac{\partial c_\alpha}{\partial \theta} \right) L_\alpha \geq 0. \tag{5.103}$$

For a one-phase decomposition, the third term in equation (5.101) leads to the condition

$$- \bar{\boldsymbol{\sigma}} \cdot \sum_{\alpha \in \mathcal{PH} \backslash 1} \dot{c}_\alpha [\![\mathbb{A}_\alpha [\bar{\boldsymbol{\varepsilon}}] - \boldsymbol{A}_\alpha^E]\!] \geq 0. \tag{5.104}$$

Thereby, the localization relation (5.62) has been used. Depending on the mechanical loading, this condition is generally not fulfilled. Therefore, if the less strict condition

$$\sum_{\alpha \in \mathcal{PH} \backslash 1} \left(\dot{c}_\alpha + \frac{\partial c_\alpha}{\partial \theta} \right) L_\alpha \geq - \bar{\boldsymbol{\sigma}} \cdot \sum_{\alpha \in \mathcal{PH} \backslash 1} \dot{c}_\alpha [\![\mathbb{A}_\alpha [\bar{\boldsymbol{\varepsilon}}] - \boldsymbol{A}_\alpha^E]\!] \tag{5.105}$$

is fulfilled during the thermomechanical process, the overall model is thermodynamically consistent. If this condition is violated, energy is withdrawn from the system due to phase transformation. In the special case of hot stamping, the assumption has been made, that the influence of the heat transfer through the boundaries is the significant quantity in the heat development in the sheet metal during the quenching process. In this situation, the thermodynamical inconsistency of the mechanical work would not have any big impact on the final state of a part.

Note that De Oliveira et al. (2010) introduced a convex macroscopic dissipation potential for the phase transformation. By introducing the thermodynamic conjugate force of the volume fraction of phase α

$$\sigma_\alpha^c = -\bar{\rho} \frac{\partial \bar{\psi}}{\partial c_\alpha}, \tag{5.106}$$

they found the convex dual macroscopic dissipation potential for the phase transformation

$$\bar{\varpi}^{*,c} = \sum_{\alpha \in \mathcal{PH}} \dot{c}_\alpha \sigma^c_\alpha \tag{5.107}$$

leading via a potential relation to the evolution equations

$$\dot{c}_\alpha = \frac{\partial \bar{\varpi}^{*,c}}{\partial \sigma^c_\alpha}. \tag{5.108}$$

The corresponding evolution equations are given by (4.98) for the JMAK, (4.104) for the KV, (4.109) for the KM, and (4.110) for the "S"-shape model.

Chapter 6

Numerical Implementation[1]

6.1 The Implementation of the Model in the ABAQUS/Standard Framework

For the simulation of the hot stamping process, the commercial implicit FE-solver ABAQUS (2016) is used. ABAQUS (2016) provides interfaces for user-defined material laws referred to as UMATHT for the thermal boundary initial value problem (BIVP), UMAT for the mechanical BIVP, and UEXPAN for the determination of the thermal expansion of the bulk.

For the numerical implementation of the thermomechanical macroscopic reference model, as well as the TMM model the separated approach is used. The thermal and mechanical balances are computed separately which means that the mechanical quantities, i.e. stress and strain, are constant in the thermal step and that the thermal quantities, i.e. temperature and volume fractions, are constant in the mechanical step. Although this leads to numerical errors in the prediction of the final thermomechanical state of a part, the numerical error is tolerable for small time steps, particularly if a simplified implementation of the thermomechanically

[1] The numerical implementation of the TMM model is discussed in the paper "Hashine-Shtrikman type mean field model for the two-scale simulation of the thermomechanical processing of steel" (Neumann and Böhlke, 2016). Two extension are introduced: the consideration of the TRIP-effect and the ULF approach for finite deformations.

coupled constitutive equations and a better convergence of the implicit FE-procedure is gained.

The flow chart of the numerical implementation of a thermomechanical constitutive law in ABAQUS/Standard is depicted in Figure 6.1. The temperature, temperature increment, and volume fractions of the

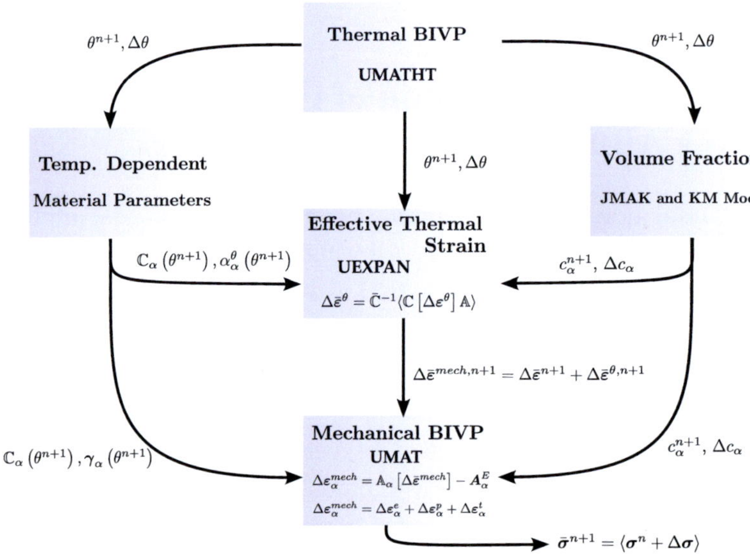

Figure 6.1: Flow chart of the numerical implementation of the multi physical model into ABAQUS/Standard

phases on the microstructure are determined in the UMATHT, where the thermal constitutive relations for the thermal conductivity and the internal energy are implemented. With the current temperature, the current mechanical parameters and the thermal strain (UEXPAN) are computed. Finally, in the UMAT, the stress response based on the mechanical constitutive relations is determined.

In the following, all quantities, represented by Ξ, at the current time step $n+1$ and at the previous time step n are abbreviated by $\Xi\left(t^{n+1}\right) = \Xi^{n+1}$ and $\Xi\left(t^{n}\right) = \Xi^{n}$, respectively. The difference of quantities at the time steps $n+1$ and n is denoted by $\Delta\Xi = \Xi^{n+1} - \Xi^{n}$. The average at the current time step is denoted by

$$\langle \Xi \rangle_{n+1} = \sum_{\alpha=1}^{N} c_{\alpha}^{n+1} \Xi_{\alpha}, \tag{6.1}$$

where $N = |\mathcal{PH}|$ is the number of phases. For evolving phases' volume fractions, in some cases, the temperature derivative of an averaged quantity is needed. The sum of all phases weighted with the temperature derivative of the volume fraction of the phases is denoted by

$$\langle \Xi \rangle_{n+1}^{\theta} = \sum_{\alpha=1}^{N} \frac{\mathrm{d}c_{\alpha}^{n+1}}{\mathrm{d}\theta^{n+1}} \Xi_{\alpha}. \tag{6.2}$$

6.2 The Heat Equation, Specific Heat Capacity, and Conductivity

The governing heat equation is the same for the macroscopic reference model (5.14) and the TMM model (5.97) and reads in time discretized form

$$\bar{\rho} \int_{\theta_0}^{\theta^{n+1}} \bar{\kappa}\, \mathrm{d}\theta - \operatorname{div}\left(\bar{\lambda}^{n+1} \boldsymbol{g}^{n+1}\right) = \sum_{\alpha=2}^{N} \Delta c_{\alpha} L_{\alpha} \tag{6.3}$$

if the specific heat capacity and the thermal conductivity of the phases are temperature dependent. In the UMATHT, the current internal energy has to be determined. With the assumption, that the volume fractions are constant, the macroscopic internal energy reads

$$\int_{\theta_0}^{\theta^{n+1}} \bar{\kappa}\,\mathrm{d}\theta = \left\langle \int_{\theta_0}^{\theta^{n+1}} \kappa\,\mathrm{d}\theta \right\rangle_{n+1}$$
$$= \left\langle e^{n+1} \right\rangle_{n+1} = \bar{e}^{n+1} = \left\langle e^{n} + \tilde{\kappa}(\theta^{n+1}) - \tilde{\kappa}(\theta^{n}) \right\rangle_{n+1}, \qquad (6.4)$$

where the primitives of the specific heat capacities has been denoted by $\tilde{\kappa}_\alpha$. Furthermore, the current macroscopic heat flux has to be determined in the UMATH. With the thermal HS localization relation (5.48) and (5.38), which also includes the simple mixture (5.15) as the upper first-order bound, the current macroscopic heat flux reads

$$\bar{\boldsymbol{q}}^{n+1} = \bar{\lambda}^{n+1}\bar{\boldsymbol{g}}^{n+1} = \left\langle \frac{\lambda^{n+1}}{\lambda^{n+1} + 2\lambda_0} \right\rangle_{n+1} \left\langle \frac{1}{\lambda^{n+1} + 2\lambda_0} \right\rangle_{n+1}^{-1} \bar{\boldsymbol{g}}^{n+1}. \qquad (6.5)$$

6.3 The Diffusionless and Diffusion-Driven Phase Transformation

In the non-isothermal case, there are two possibilities of implementing the JMAK equation. The first one is the discretization of the rate form of the JMAK model (4.100) directly derived from Scheil's additivity rule. Applying the implicit Euler method, the equation for the current volume fraction reads

$$c_\alpha^{n+1} = c_\alpha^{n} - \Delta t \frac{\left(\int_0^{t^{n+1}} - \frac{\tau_\alpha(\theta(\xi), c_\alpha^{n+1})^{-1 - a_\alpha(\theta(\xi))}}{a_\alpha(\theta(\xi)) b_\alpha(\theta(\xi)) \left(1 - c_\alpha^{n+1}\right)} \,\mathrm{d}\xi \right)^{-1}}{\tau_\alpha\left(\theta^{n+1}, c_\alpha^{n+1}\right)}, \qquad (6.6)$$

where $\tau_\alpha(\theta, c_\alpha)$ is given by equation (4.99). Note, that the evaluation of this equation requires a great deal of effort. Due to the fact that the integrand has to be evaluated at the current volume fraction c_α^{n+1}, both the integration has to be performed in each step from zero to the current time t^{n+1}, and the temperature history has to be saved for each material point of the bulk.

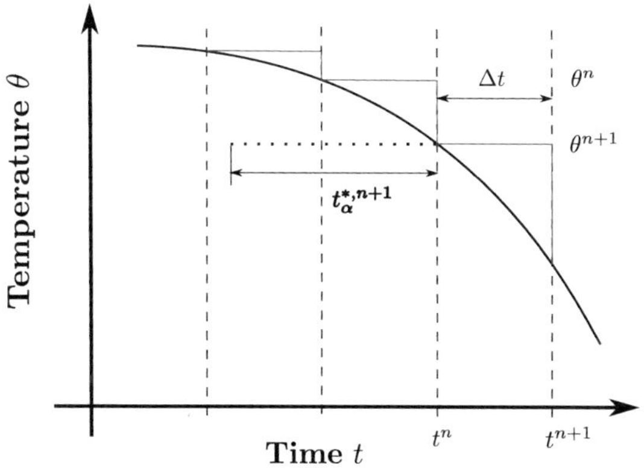

Figure 6.2: The isothermal time step method and depiction of the fictive isothermal process and time

The second way to implement the JMAK model in the non-isothermal case is to use the isothermal time step method. At the current time, the current temperature θ^{n+1} and the volume fraction at the beginning of the time step c_α^n are given. The idea is to pretend that the process to reach the volume fraction c_α^n was isothermal. This process is referred to as a fictitious process. The time that it takes in this fictitious process to reach a volume fraction of c_α^n at the temperature level θ^{n+1}, is denoted by $t_\alpha^{*,n+1}$ and is determinable by equation (4.99)

$$t_\alpha^{*,n+1} = \tau_\alpha\left(\theta^{n+1}, c_\alpha^n\right) = \left(-\frac{\ln\left(1 - \frac{c_\alpha^n}{c_\alpha^r}\right)}{b_\alpha\left(\theta^{n+1}\right)}\right)^{\frac{1}{a_\alpha\left(\theta^{n+1}\right)}} . \tag{6.7}$$

Figure 6.2 depicts the fictitious process and the fictive time $t_\alpha^{*,n+1}$. The current volume fraction is obtained by the JMAK equation applied for the isothermal time step

$$c_\alpha^{n+1} = c_\alpha^r \left(1 - \exp\left(-b_\alpha\left(\theta^{n+1}\right)\left(t_\alpha^{*,n+1} + \Delta t\right)^{a_\alpha\left(\theta^{n+1}\right)}\right)\right). \tag{6.8}$$

The isothermal time step method is an efficient numerical implementation of the JMAK equation, but is highly time step dependent.

Using the implicit Euler method, the time discretization of the KV model (4.104) leads to an implicit nonlinear equation for the determination of the current volume fraction

$$g_\alpha(c_\alpha^{n+1}) = c_\alpha^{n+1} - c_\alpha^n + \Delta t f_\alpha^S(S) f_\alpha^\theta(\theta^{n+1}) f_\alpha^C(C) f_\alpha^c(c_\alpha^{n+1}) f_\alpha^r\left(\frac{\dot{\Delta\theta}}{\Delta t}\right) = 0, \tag{6.9}$$

with the influence functions given in (4.94), (4.95), and (4.105). Applying Newton's method, the iteration for the determination of the volume fraction reads

$$c_\alpha^{n+1,i+1} = c_\alpha^{n+1,i} - \frac{g_\alpha(c_\alpha^{n+1,i})}{g'_\alpha(c_\alpha^{n+1,i})}, \tag{6.10}$$

where i is the iteration index. The derivatives are given by

$$g'_\alpha = 1 + \Delta t f_\alpha^S(S) f_\alpha^\theta(\theta^{n+1}) f_\alpha^C(C) \frac{\mathrm{d} f_\alpha^c(c_\alpha^{n+1})}{\mathrm{d} c_\alpha^{n+1}} f_\alpha^r\left(\frac{\dot{\Delta\theta}}{\Delta t}\right) \tag{6.11}$$

$$\frac{\mathrm{d} f_\alpha^c(c_\alpha^{n+1})}{\mathrm{d} c_\alpha^{n+1}} = n_\alpha f_\alpha^c(c_\alpha^{n+1}) \left(\frac{1}{c_\alpha^{n+1}} + \ln\left(\frac{1 - c_\alpha^{n+1}}{c_\alpha^{n+1}}\right) + \frac{c_\alpha^{n+1}}{1 - c_\alpha^{n+1}} - 1\right). \tag{6.12}$$

Due to the iteration, the KV is costlier than the isothermal time step method of the JMAK equation, however, it is less time step dependent.

For the description of the diffusionless phase transformation, the rate form of the Koistinen-Marburger model (4.109) is used. Discretizing the rate law by use of the implicit Euler scheme, in the case of cooling and a current temperature below the martensite formation start temperature, the current volume fraction of the martensitic phase is given by

$$c_m^{n+1} = \frac{c_m^n + \gamma\left(\theta^{n+1} - \theta^n\right) c_m^r}{1 + \gamma\left(\theta^{n+1} - \theta^n\right)}. \tag{6.13}$$

An equivalent procedure leads to the "S"-shape model (4.110) in discretized form

$$c_m^{n+1} - c_m^n - b\Delta\theta\left(c_m^{n+1} - c_m^r\right)\frac{c_m^{n+1}}{c_m^r} = 0, \tag{6.14}$$

with the physically reasonable solution of the quadratic equation

$$c_m^{n+1} = \varsigma + \sqrt{\varsigma^2 - \frac{c_m^r c_m^n}{b\Delta\theta}}, \qquad \varsigma = \frac{1}{2}\left(\frac{c_m^r}{b\Delta\theta} + c_m^r\right). \tag{6.15}$$

6.4 The Update Lagrangian Formulation for the Description of Finite Deformations

In the FE implementation of the mechanical BIVPs, there are two possibilities of formulating the principle of virtual work: in terms of the reference placement and in terms of the current placement. The linearization of the latter form leads to the updated Lagrangian formulation (ULF) (see, e.g., Wriggers, 2008; Kim, 2015). Thereby, the current placement of the body $\chi_{t^n}(\mathcal{B})$ is used as a new reference placement and the incremental placement $\chi_{t^{n+1}}(\mathcal{B})$ as the incrementally deformed state. Introducing the mapping between the new reference placement and the incremental placement $\chi_{R,t^n}(\chi_{t^n}(\mathcal{B}), t^{n+1})$, Dixit and Dixit (2008) introduced kinematic quantities corresponding to the ones introduced in Chapter 3, which describe the motion of particles between these placements. They showed, that, in the special case of incremental small deformations, the Green's strain increment (3.16) can be decomposed in an elastic and inelastic contribution. This leads to the small incremental deformation approximation via the infinitesimal strain increment

$$\Delta\bar{\varepsilon} = \Delta\bar{\varepsilon}^e + \Delta\bar{\varepsilon}^i. \tag{6.16}$$

In ABAQUS/Standard, the infinitesimal strain increment is approximated via the material Hencky strain increment (3.15) (see, e.g., ABAQUS, 2016) with the series expansion

$$\Delta\bar{\varepsilon} = \ln\left(\Delta\bar{\boldsymbol{U}}\right) = \frac{1}{2}\left(\Delta\bar{\boldsymbol{F}}^{\mathrm{T}}\Delta\bar{\boldsymbol{F}} - \boldsymbol{I}\right) - \frac{1}{4}\left(\Delta\bar{\boldsymbol{F}}^{\mathrm{T}}\Delta\bar{\boldsymbol{F}} - \boldsymbol{I}\right)^2 + \ldots, \tag{6.17}$$

where the increment of the deformation gradient and the increment of the right Cauchy-Green tensor read

$$\Delta\bar{\boldsymbol{F}} = \bar{\boldsymbol{F}}^{n+1}(\bar{\boldsymbol{F}}^{n})^{-1}, \quad \Delta\bar{\boldsymbol{C}} = \Delta\bar{\boldsymbol{F}}^{\mathrm{T}}\Delta\bar{\boldsymbol{F}}, \tag{6.18}$$

respectively. Note, that the consideration of the spatial Hencky strain increment (3.14) leads to an equivalent result. Furthermore, note, that this infinitesimal strain increment is objective, which can be seen if only rigid body motion is considered.

6.5 The Phenomenological Reference Model

Since the ULF is used for the description of finite deformations and the separated implementation method is used, the mechanical constitutive law has to be given in rate form

$$\Delta\bar{\boldsymbol{\sigma}} = \bar{\mathbb{C}}^{n+1}\left[\Delta\bar{\varepsilon} - \Delta\bar{\varepsilon}^p - \Delta\bar{\varepsilon}^{tr} - \Delta\bar{\varepsilon}^t - \bar{\alpha}\Delta\bar{\theta}\boldsymbol{I}\right], \tag{6.19}$$

where the current effective stiffness tensor and the current effective thermal expansion coefficient are given by

$$\bar{\mathbb{C}}^{n+1} = 3 \left\langle K(\theta^{n+1} \right\rangle_{n+1} \mathbb{P}^{1,\,\text{iso}} + 2 \left\langle G(\theta^{n+1}) \right\rangle_{n+1} \mathbb{P}^{2,\,\text{iso}}, \tag{6.20}$$

and $\bar{\alpha} = \langle \alpha \rangle_{n+1}$, respectively.

Note, that due to the separated implementation of the thermomechanical constitutive law, the temperature and the volume fractions are constant in the mechanical increment. Due to this, the increment of the stiffness tensor can be neglected in the rate form of the mechanical constitutive law.

With the isotropic elasticity, the stress increment can be decomposed into a spherical and a deviatoric part

$$\Delta\bar{\boldsymbol{\sigma}}^{\circ} = 3 \left\langle K(\theta^{n+1}) \right\rangle_{n+1} \left(\Delta\bar{\varepsilon}^{\circ} - \Delta\bar{\varepsilon}^{t} - \bar{\alpha}\Delta\bar{\theta}\boldsymbol{I} \right), \tag{6.21}$$

$$\Delta\bar{\boldsymbol{\sigma}}' = 2 \left\langle G(\theta^{n+1}) \right\rangle_{n+1} \left(\Delta\bar{\varepsilon}' - \Delta\bar{\varepsilon}^{p} - \Delta\bar{\varepsilon}^{tr} \right). \tag{6.22}$$

Since the macroscopic TRIP strain (5.6) is dependent on the deviatoric part of the macroscopic stress, the latter equation can be simplified to

$$\bar{\boldsymbol{\sigma}}^{n+1\,\prime} = \frac{\bar{\boldsymbol{\sigma}}^{n\,\prime}}{\gamma^{tr,n+1}} + 2 \frac{\left\langle G(\theta^{n+1}) \right\rangle_{n+1}}{\gamma^{tr,n+1}} \left(\Delta\bar{\varepsilon}' - \Delta\bar{\varepsilon}^{p} \right], \tag{6.23}$$

where

$$\gamma^{tr,n+1} = 1 + \frac{3}{2} \sum_{\alpha=1}^{N} k_{\alpha}^{tr} \phi'(c_{\alpha}^{n+1}) \Delta c_{\alpha} \mathcal{H}(\Delta c_{\alpha}) \tag{6.24}$$

has been introduced. Since $\gamma^{tr,n+1} \geq 1$ holds, the TRIP mechanism appears as a shear softening effect (cf. Chapter 5.1). Based on this expression for the deviatoric stress, the plasticity is determined by introducing an elastic deviatoric trial stress and by radial return mapping of the stress state on the yield surface (see, e.g., Simo and Hughes, 1998)

$$\bar{\phi}\left(\bar{\boldsymbol{\sigma}}^{n+1}, \bar{\varepsilon}^{p,n+1}\right) = \| \, \bar{\boldsymbol{\sigma}}^{n+1\,\prime} \, \| - \sqrt{\frac{2}{3}} \left\langle s^{f,n+1} \left(\varepsilon^{0,n+1} + \bar{\varepsilon}^{p,n+1}\right)^{s^{e,n+1}} \right\rangle_{n+1} = 0. \tag{6.25}$$

where the current Swift parameters are defined by

$$s^{f,n+1} = s^{f,0} + s^{f,1}\theta^{n+1} \tag{6.26}$$

$$s^{e,n+1} = s^{e,0} + s^{e,1}\theta^{n+1}. \tag{6.27}$$

The transformation strain increment is obtained by deriving equation (5.7) with respect to time and a time discretization

$$\Delta\bar{\varepsilon}^{t} = \left\langle \zeta \Delta c \delta^{n}\left(c^{n+t}\right)\right\rangle_{n+1}. \tag{6.28}$$

Thereby, the Dirac delta function has been numerically realized by

$$\delta^{n}(c_{\alpha}^{n+1}) = \begin{cases} \frac{1}{c_{\alpha}^{n+1}} & \text{if} \quad c_{\alpha}^{n} = 0 \text{ and} \Delta c_{\alpha} \neq 0 \\ 0 & \text{else} \end{cases}. \tag{6.29}$$

In ABAQUS/Standard the algorithmic tangent is needed. For the macroscopic reference model, the tangent reads (see, e.g., Simo and Hughes, 1998)

$$\bar{\mathbb{C}}^{\,\text{alg}} = \begin{cases} 3\left\langle K^{n+1}\right\rangle_{n+1} \mathbb{P}^{1,\,\text{iso}} + 2\tilde{G}^{n+1}\mathbb{P}^{2,\,\text{iso}} & \text{if the step is elastic} \\ 3\left\langle K^{n+1}\right\rangle_{n+1} \mathbb{P}^{1,\,\text{iso}} + 2\tilde{G}^{n+1}\tilde{\mu}^{n+1}\mathbb{P}^{2,\,\text{iso}} - \\ \qquad 2\tilde{\mu}^{n+1}\tilde{G}^{n+1}\bar{\boldsymbol{N}}^{n+1} \otimes \bar{\boldsymbol{N}}^{n+1} & \text{else} \end{cases}, \tag{6.30}$$

where

$$\tilde{G}^{n+1} = \frac{\left\langle G^{n+1}\right\rangle_{n+1}}{\gamma^{tr,n+1}}, \qquad \bar{\boldsymbol{N}}^{n+1} = \frac{\bar{\boldsymbol{\sigma}}^{n+1,\,\text{tr}}}{\|\bar{\boldsymbol{\sigma}}^{n+1,\,\text{tr}}\|}, \tag{6.31}$$

and

$$\bar{\mu}^{n+1} = 1 - \frac{2\tilde{G}^{n+1}\Delta\bar{\gamma}}{\|\bar{\boldsymbol{\sigma}}^{n+1,\,\mathrm{tr}}\|}, \qquad \tilde{\mu}^{n+1} = \frac{1}{1 + \frac{\langle \iota^{p,n+1\,\prime} \rangle_{n+1}}{3\tilde{G}^{n+1}}} - \left(1 - \bar{\mu}^{n+1}\right) \quad (6.32)$$

has been introduced. Thereby, the elastic trial stress is denoted by $\bar{\boldsymbol{\sigma}}^{n+1,\,\mathrm{tr}}$ and the increment of the consistency parameter by $\Delta\bar{\gamma}$.

6.6 The Material Behavior of Phases in the TMM Model

The ULF is also used in the TMM model-based simulations of the hot stamping process. Due to this, the mechanical constitutive law for each phase has to be given in rate form

$$\Delta\boldsymbol{\sigma}_\alpha = \mathbb{C}_\alpha^{n+1}\left[\Delta\boldsymbol{\varepsilon}_\alpha - \Delta\boldsymbol{\varepsilon}_\alpha^{p} - \Delta\boldsymbol{\varepsilon}_\alpha^{tr} - \Delta\boldsymbol{\varepsilon}_\alpha^{t} - \alpha_\alpha\Delta\theta\boldsymbol{I}\right], \quad (6.33)$$

whereby, the current isotropic stiffness tensor reads

$$\mathbb{C}_\alpha^{n+1} = 3K_\alpha^{n+1}\mathbb{P}^{1,\,\mathrm{iso}} + 2G_\alpha^{n+1}\mathbb{P}^{2,\,\mathrm{iso}}. \quad (6.34)$$

To avoid further internal iterations on the material behavior level, the increment of the TRIP strain of the phases is approximated with the macroscopic stress at the beginning of the step

$$\Delta\boldsymbol{\varepsilon}_a^{tr} = \frac{3}{2}\sum_{\alpha=2}^{N} k_\alpha^{tr}\phi_\alpha(c_\alpha^{n+1})\Delta c_\alpha\mathcal{H}(\Delta c_\alpha)\,\bar{\boldsymbol{\sigma}}^{n\,\prime}. \quad (6.35)$$

Due to the isothermal step method of the JMAK equation implementation in particular the phase transformation needs to be simulated with small time increments, this approximation leads to small errors and a less complex numerical implementation.

The plasticity is determined by introducing an elastic deviatoric trial phase stress and by radial return mapping of the phase stress state on the phase yield surface (see, e.g., Simo and Hughes, 1998).

The phases' transformation strain increment is obtained by deriving equation (5.35) with respect to time and a time discretization

$$\Delta\varepsilon_\alpha^t = \zeta_\alpha \Delta c_\alpha \delta^n(c_\alpha^{n+t}), \tag{6.36}$$

with the numerical realization of the Dirac delta function (6.29).

6.7 The Strain Rate Localization Relation

Due to the ULF, only phase strain increments are available in a time step. For this reason, an incremental form of the phase interaction law for the scale transition has to be found. Assuming a constant stiffness of the homogeneous comparison material, the derivation of the phase interaction law (5.60) with respect to time yields

$$\dot{\boldsymbol{\sigma}}_\alpha - \langle\dot{\boldsymbol{\sigma}}\rangle - \sum_{\alpha=1}^{N} \dot{c}_\alpha \boldsymbol{\sigma}_\alpha = \mathbb{L}[\dot{\varepsilon}_\alpha - \dot{\bar{\varepsilon}}], \tag{6.37}$$

where the time derivative of a volume averaged quantity (3.41) has been used. If the macroscopic strain increment is given, the time discretization of this law reads

$$\Delta\boldsymbol{\sigma}_\alpha - \langle\Delta\boldsymbol{\sigma}\rangle_{n+1} - \sum_{\alpha=1}^{N} \Delta c_\alpha \boldsymbol{\sigma}_\alpha^{n+1} = \mathbb{L}[\Delta\varepsilon_\alpha - \Delta\bar{\varepsilon}], \tag{6.38}$$

which can can be simplified in a separated implementation approach of the thermal and mechanical BIVPs to

$$\Delta\boldsymbol{\sigma}_\alpha - \langle\Delta\boldsymbol{\sigma}\rangle_{n+1} = \mathbb{L}[\Delta\varepsilon_\alpha - \Delta\bar{\varepsilon}]. \tag{6.39}$$

This form complies with the original phase interaction law (5.60), whereby the strain rate localization relation in reduced form for isotropic elastic phases

$$\Delta\varepsilon_\alpha^\circ = \frac{\tau_\alpha^{n+1}}{K_\alpha^{n+1}} \left\langle \frac{\tau^{n+1}}{K^{n+1}} \right\rangle_{n+1}^{-1} \Bigg(\Delta\bar{\varepsilon}^\circ + K_\alpha^{n+1} \left\langle \frac{\tau^{n+1}}{K^{n+1}} \right\rangle_{n+1} \Delta\varepsilon_\alpha^{E\circ} - \left\langle \tau^{n+1} \Delta\varepsilon^{E\circ} \right\rangle_{n+1} \Bigg), \tag{6.40}$$

$$\Delta\varepsilon_\alpha' = \frac{\varsigma_\alpha^{n+1}}{G_\alpha^{n+1}} \left\langle \frac{\varsigma^{n+1}}{G^{n+1}} \right\rangle_{n+1}^{-1} \Bigg(\Delta\bar{\varepsilon}' + G_\alpha^{n+1} \left\langle \frac{\varsigma^{n+1}}{G^{n+1}} \right\rangle_{n+1} \Delta\varepsilon_\alpha^{E\,\prime} - \left\langle \varsigma^{n+1}\, \Delta\varepsilon^{E\,\prime} \right\rangle_{n+1} \Bigg) \tag{6.41}$$

can be found. Thereby, the localization parameters read

$$\tau_\alpha^{n+1} = \frac{3K_\alpha^{n+1}}{3K_0 + 4G_0 + 3\delta K_\alpha^{n+1}}, \qquad \varsigma_\alpha^{n+1} = \frac{2G_\alpha^{n+1}}{\frac{5G_0(3K_0+6G_0)}{3K_0+4G_0} + 2\delta G_\alpha^{n+1}}, \tag{6.42}$$

with the definitions $\delta K_\alpha^{n+1} = K_\alpha^{n+1} - K_0$ and $\delta G_\alpha^{n+1} = G_\alpha^{n+1} - G_0$. Note, that depending on the constant temperature and volume fraction in the mechanical step, the localization parameters change from step to step. Due to this, the localization parameters have to be determined in each step.

Note, that due to the dependence of the phases' eigenstrain tensor (in particular the plastic strain) on the phase strain the deviatoric part of the strain localization relation (6.41) is a set of N coupled nonlinear tensor valued tensor equations. For the determination of the phase strains, an iterative scheme as Newton's or Banach's fix-point method has to be used.

The macroscopic algorithmic tangent is needed for the implementation in the implicit solver ABAQUS/Standard. Due to the ULF and implementation of the phase transformation laws, the time steps have to be small compared to the transformation time. The effective algorithmic consistent tangent can be computed by (see, e.g., Jöchen, 2013; Neumann and Böhlke, 2016)

$$\bar{\mathbb{C}}^{alg} = \frac{\partial \Delta \bar{\boldsymbol{\sigma}}^{n+1}}{\partial \Delta \bar{\varepsilon}^{n+1}} = \left\langle \mathbb{C}^{algo} \left(\mathbb{A} - \frac{\partial \boldsymbol{A}^{E}}{\partial \bar{\varepsilon}^{n+1}} \right) \right\rangle \approx \left\langle \mathbb{C}^{alg} \mathbb{A} \right\rangle, \tag{6.43}$$

with the algorithmic tangent of phase α $\mathbb{C}_{\alpha}^{alg} = \partial \Delta \boldsymbol{\sigma}_{\alpha} / \partial \Delta \varepsilon_{\alpha}$ given in, e.g., Simo and Hughes (1998).

Chapter 7

Material Parameter Identification[1]

7.1 Thermomechanical Parameters of 42CrMo4

7.1.1 Bulk Behavior Concerning Parameters

Besides the steel grade 22MnB5, which is usually used for the hot stamping process, the steel 42CrMo4 is considered. The latter steel grade is used for the investigation of the latent heat released during phase transformation and the transformation strain on the final stress and strain state of sheet metal. The following parameters of the steel 42CrMo4 for the underlying model have been used in Neumann and Böhlke (2016) and are listed here for reasons of consistency.

Table 7.1: Chemical composition of 42CrMo4 in percent by weight (see, e.g., Schwenk et al., 2012; Miokovic, 2005)

C	Si	Mn	P	S	Cr	Ni	Mo	Sn	Al
0.425	0.309	0.702	0.019	0.013	1.014	0.098	0.198	0.013	0.021

The chemical composition of the steel grade 42CrMo4 is given in Table 7.1, respectively.

[1] The material parameter identification of the steel grade 42CrMo4 is shown in the paper "Hashine-Shtrikman type mean field model for the two-scale simulation of the thermomechanical processing of steel" (Neumann and Böhlke, 2016). For the sake of completeness the results are represented.

Table 7.2: Young's moduli, Poisson's ratios, thermal expansion coefficients, transformation strains, specific heat capacities, and heat conductivities of phases of 42CrMo4 (Miokovic, 2005; Schwenk et al., 2012)

	Austenite	Ferrite/Pearlite	Bainite	Martensite
E_α: m_α^E $\left[\frac{10^8\text{Pa}}{\text{K}}\right]$	-1.63	-1.0	-0.8276	-0.73
n_α^E $\left[10^{11}\text{Pa}\right]$	2.55	2.51	2.41	2.36
ν_α: m_α^ν $\left[\frac{10^{-5}}{\text{K}}\right]$	7.0	5.7	5.1	4.74
n_α^ν $[-]$	0.253	0.265	0.267	0.266
α_α $\left[\frac{10^{-5}}{\text{K}}\right]$	2.326	1.45	1.6	1.24
ζ_α $\left[10^{-3}\right]$	-3.0	3.0	3.0	5.0
κ_α $\left[\frac{\text{J}}{\text{kgK}}\right]$	572.07	662.12	541.72	546.73
Λ_α: m_α^Λ $\left[\frac{10^{-4}\text{W}}{\text{mK}^2}\right]$	224.55	-260.13	-189.19	-118.02
n_α^Λ $\left[\frac{10^{-1}\text{W}}{\text{mK}}\right]$	0.0	520.74	455.63	376.88

For the elasticity of both steel grades, the linear temperature dependence of the Young's modulus and the Poisson's ratio

$$E_\alpha(\theta) = m_\alpha^E\theta + n_\alpha^E, \quad \nu_\alpha(\theta) = m_\alpha^\nu\theta + n_\alpha^\nu, \quad \Lambda_\alpha(\theta) = m_\alpha^\Lambda\theta + n_\alpha^\Lambda \tag{7.1}$$

is chosen. The fitting is done by a least square fit of the linear functions to the experimental results $f_{\alpha,i}$

$$\tilde{\boldsymbol{v}}_\alpha^f = \begin{pmatrix} \tilde{m}_\alpha^f \\ \tilde{n}_\alpha^f \end{pmatrix} = \arg\left(\min_{\boldsymbol{v}_\alpha^f}\left(\sum_{i=1}^{N_{exp}}\left(f_\alpha\left(\theta_i, \boldsymbol{v}_\alpha^f\right) - f_{\alpha,i}\right)^2\right)\right), \; f \in \{E, \nu, \Lambda\}, \tag{7.2}$$

where N_{exp} denotes the number of experimentally determined values (see Figure 7.1). Note, that all fits are created with the optimization software for nonlinear problems MIDACO described in Schlueter et al. (2012).

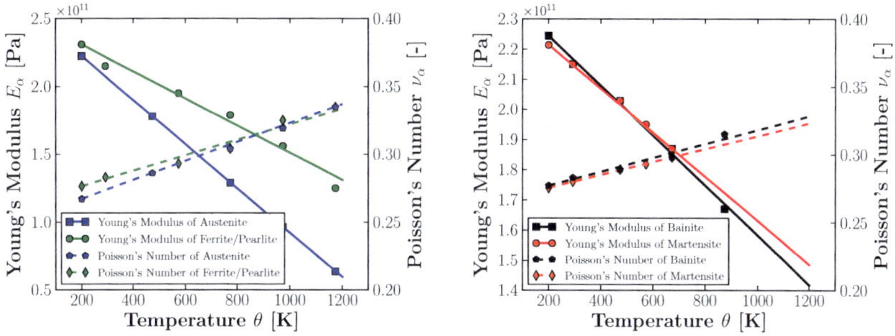

Figure 7.1: Temperature dependence of the Young's moduli, and Poisson's numbers of considered phases (dots: experimental results; lines: fits)

For 42CrMo4 the elasticity and conductivity parameters are given in Table 7.2, where the constant isotropic thermal expansion coefficients are also listed.

The hardening behavior of the different phases is dependent on both the temperature and the accumulated plastic strain (see Figure 7.2, where $\Phi_\alpha = \sigma^0_\alpha + \partial\psi^i_\alpha/\partial\varepsilon^p_\alpha$ has been introduced). By use of the least square fit method

$$\tilde{\boldsymbol{\beta}}_\alpha(\tilde{\theta}) = \arg\left(\min_{\boldsymbol{\beta}_\alpha}\left(\sum_{i=1}^{N_{exp}}(f_\alpha(\tilde{\theta},\boldsymbol{\beta}_\alpha,\varepsilon^p_{\alpha,i}) - f^{\tilde{\theta}}_{\alpha,i})^2\right)\right), \tag{7.3}$$

where the vector $\beta_\alpha = \left(\sigma^f_\alpha\ \sigma^\infty_\alpha\ \theta^0_\alpha\ \theta^\infty_\alpha\right)^{\mathrm{T}}$ has been defined, the Voce law with the hardening potential given in equation (4.44) is fitted to the associated experimentally determined curves at a certain temperature level $\tilde{\theta}$. This procedure is done for each temperature level which has been measured for the phases. For the determination of the hardening behavior of the phases at an arbitrary temperature, the Voce parameters σ^f_α, σ^∞_α, θ^0_α, and θ^∞_α, which are listed in Table 7.3, are interpolated linearly between the two corresponding temperature levels.

Table 7.3: Temperature-dependent hardening parameters of the Voce hardening law of 42CrMo4 (Neumann and Böhlke, 2016)

Austenite θ [K]	473.15	673.15	873.15	1173.15		
σ_a^f $[10^8\text{Pa}]$	2.5	1.8	1.5	0.53		
σ_a^∞ $[10^8\text{Pa}]$	2.9	2.25	1.55	0.7		
θ_a^0 $[10^{10}\text{Pa}]$	2.7	1.2	1.0	0.7		
θ_a^∞ $[10^9\text{Pa}]$	2.3	1.6	1.5	0.4		

Ferrite θ [K]	293.15	373.15	573.15	773.15	973.15	1173.15
σ_f^f $[10^8\text{Pa}]$	8.85	7.5	6.0	4.6	1.6	1.0
σ_f^∞ $[10^8\text{Pa}]$	10.1	9.9	9.3	6.5	3.15	0.01
θ_f^0 $[10^{10}\text{Pa}]$	26.0	9.3	8.7	8.5	6.8	$1.0 \cdot 10^{-4}$
θ_f^∞ $[10^9\text{Pa}]$	3.4	2.5	1.8	1.8	1.7	$1.0 \cdot 10^{-3}$

Bainite θ [K]	293.15	473.15	673.15	873.15		
σ_b^f $[10^8\text{Pa}]$	9.0	8.5	8.0	3.0		
σ_b^∞ $[10^9\text{Pa}]$	1.25	1.1	1.0	0.45		
θ_b^0 $[10^{10}\text{Pa}]$	5.0	5.0	2.3	2.0		
θ_b^∞ $[10^8\text{Pa}]$	8.0	8.8	6.0	5.0		

Martensite θ [K]	293.15	373.15	473.15	573.15		
σ_m^f $[10^9\text{Pa}]$	1.62	1.59	1.5	1.4	1.04	
σ_m^∞ $[10^9\text{Pa}]$	1.94	1.74	1.64	1.5	0.0	
θ_m^0 $[10^{10}\text{Pa}]$	26.0	9.0	4.63	3.71	0.0	
θ_m^∞ $[10^9\text{Pa}]$	14.0	12.9	8.43	5.74	0.0	

7.1.2 Phase Transformation Concerning Parameters

The transformation start temperatures Θ_α^s for austenite in the heating case, and ferrite/pearlite, bainite, and martensite in the cooling case are shown in Table 7.4.

The phase transformation accompanying transformation strain is determined from a dilatation test. Therefore, the transformation strain is related to a reference temperature. Choosing the initial temperature ($\theta_{ref} = 273.15\text{K}$) as the reference temperature, the transformation strain of ferrite/pearlite to austenite during heating and from austenite to martensite during quenching read $\zeta_a = -0.919$ and $\zeta_m = 0.954$, respectively (see, e.g., Schwenk et al., 2012). In this work, the transformation strain is related to the respective transformation finish temperatures of the arising phases (see Figure 7.3). For the transformation strain from ferrite/pearlite to austenite, and from austenite to martensite, one obtains $\zeta_a = -0.003$, and $\zeta_m = 0.005$, respectively. The transformation strain of bainite is set to be $\zeta_b = 0.003$.

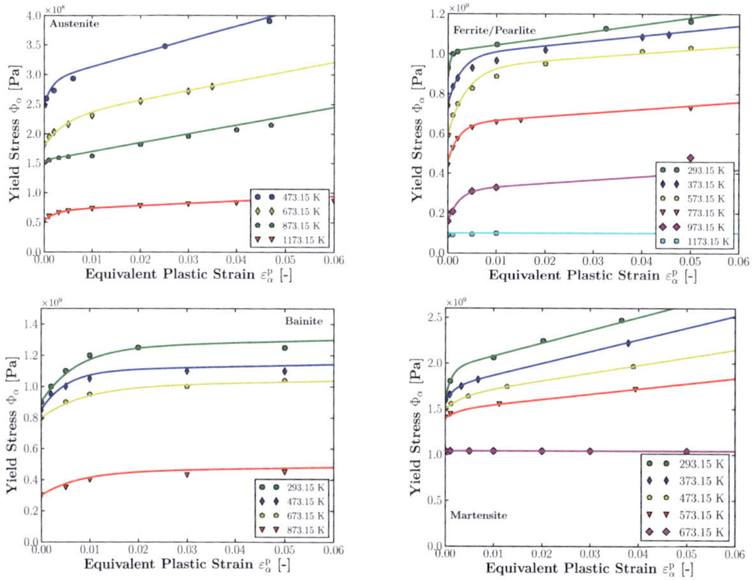

Figure 7.2: Temperature dependence of the yield behavior of considered phases of the steel grade 42CrMo4 (dots: experimental results; lines: fits)

The latent heats, which are released during phase transformation from austenite to ferrite/pearlite L_{fp}, from austenite to bainite L_b, and from austenite to martensite L_m, and the spent energy, which is consumed during the transformation from ferrite/pearlite to austenite L_a, are given in Table 7.4.

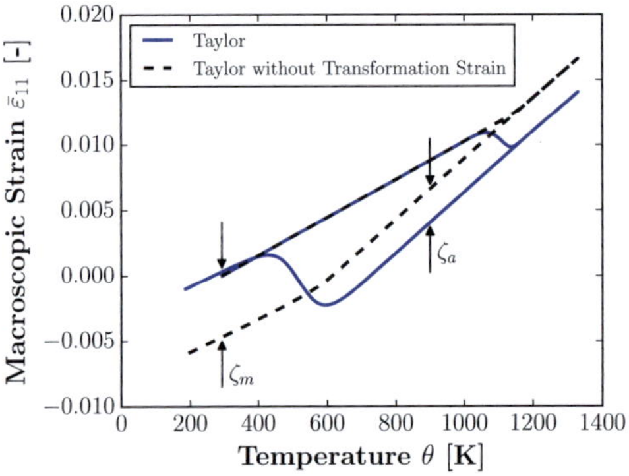

Figure 7.3: Definition of the transition strain in an incrementally implemented constitutive law

Diffusion-Driven Phase Transformation

The diffusion-driven phase transformation parameters are determined by use of the TTT diagram. In the TTT diagram, the isolines of the transformation start $c_\alpha = 0.01$ and transformation end $c_\alpha = 0.99$ are depicted in the temperature-time diagram. For a certain temperature level $\tilde{\theta}$, the transformation start time $t_s(\tilde{\theta})$ and the transformation end time $t_e(\tilde{\theta})$ in an isothermal process can be found as shown in Figure 7.4. In this isothermal condition, the JMAK model (4.75) can be used to set up two equations for the determination of the two Avrami parameters $a_\alpha(\tilde{\theta})$ and $b_\alpha(\tilde{\theta})$ (see, e.g., Hömberg, 1996)

$$0.99 = \exp\left(-b_\alpha\left(\tilde{\theta}\right) t_s\left(\tilde{\theta}\right)^{a_\alpha(\tilde{\theta})}\right), \quad 0.01 = \exp\left(-b_\alpha\left(\tilde{\theta}\right) t_e\left(\tilde{\theta}\right)^{a_\alpha(\tilde{\theta})}\right), \tag{7.4}$$

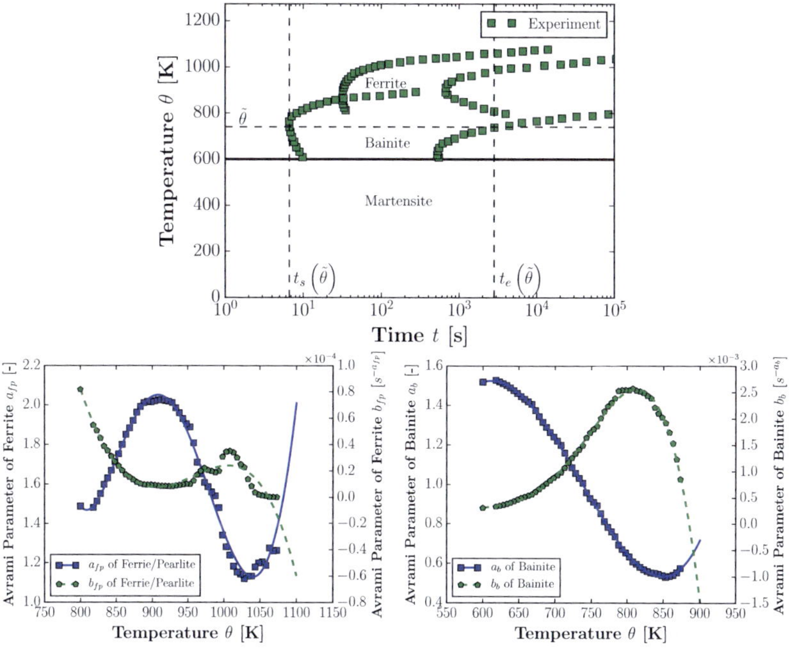

Figure 7.4: Determination of $t_s(\tilde{\theta})$ and $t_e(\tilde{\theta})$ from the TTT diagram of 42CrMo4 (left) and Avrami parameters of ferrite/pearlite and bainite with polynomial fit (right)

with the solutions

$$a_\alpha\left(\tilde{\theta}\right) = \frac{\ln\left(\ln\left(0.01\right) - \ln\left(0.99\right)\right)}{\ln\left(t_e\left(\tilde{\theta}\right)\right) - \ln\left(t_s\left(\tilde{\theta}\right)\right)}, \; b_\alpha\left(\tilde{\theta}\right) = -\ln\left(0.99\right) t_e\left(\tilde{\theta}\right)^{-a_\alpha\left(\tilde{\theta}\right)}. \tag{7.5}$$

On the right-hand side of Figure 7.4, the Avrami parameters of the phases ferrite/pearlite and bainite with the polynomial interpolation (4.79) and (4.78) are depicted.

Diffusionless Phase Transformation

The parameter γ in the Koistinen-Marburger model (4.107) can be identified assuming a purely austenitic-martensitic phase transformation. In this case, the retained austenite is one $c_m^r = 1$. If the martensite formation finish temperature M_f is known ($c_m(\theta = M_f) = c_{mf}$, γ is given by $\gamma = \ln(1 - c_{mf})/(M_s - M_f)$. With the martensite formation finish temperature $M_f = 273.15\text{K}$ for the steel 42CrMo4 and $c_m(\theta = M_f) = 0.99$, the transformation parameter reads $\gamma = -0.01354\text{K}^{-1}$. A similar approach for the "S"-shape model (4.108) leads to

$$a = \frac{1 - c_{ms}}{c_{ms}}, \qquad b = \frac{1}{M_f - M_s} \ln\left(\frac{c_{ms} - c_{mf}c_{ms}}{c_{mf} - c_{mf}c_{ms}}\right), \tag{7.6}$$

where c_{ms} denotes the volume fraction of martensite at the martensite start temperature. Setting $c_{ms} = 0.01$ and $c_{mf} = 0.99$, the parameters of the "S"-shape model read $a = 99.0$ and $b = 0.028688121\text{K}^{-1}$, respectively.

Another approach for determining the diffusionless transformation parameters of both models is to use the least square fitting method. The resulting dilatation curve of the thermomechanical simulation is compared with the experimental one and the error squares are minimized. The resulting parameters of such an approach are $a_{\text{opt}} = 40.79$, $b_{opt} = 0.030346\text{K}^{-1}$, and $c_{\text{opt}} = 640.17\text{K}$ for the "S"-shape model. On the right-hand side of Figure 8.3, the different resulting dilatation curves with the best approximation gained from the least square method are depicted. Subsequently, these optimized parameters are used in further numerical considerations.

All parameters concerning the phase transformation are listed in Table 7.4.

7.2 Thermomechanical Parameters of 22MnB5

7.2.1 Bulk Behavior Concerning Parameters

Due to lack of experimental results, Young's modulus of the phases of 22MnB5 has been chosen to be the same as that of 42CrMo4. To get a

Table 7.4: Phase transformation parameters, transformation temperatures, and latent heat of 42CrMo4 (due to brevity, the units of most parameters are omitted; since SI units are used they can be easily determined from the corresponding equations)

Transformation Parameters of the Diffusion-Driven Phase Transformation					
	$a_\alpha^{a/b}$	$b_\alpha^{a/b}$	$c_\alpha^{a/b}$	$d_\alpha^{a/b}$	$e_\alpha^{a/b}$
Ferrite					
$a_f(\theta)$	$8.22 \cdot 10^{3}$	$-4.21 \cdot 10^{1}$	$8.52 \cdot 10^{-2}$	$-8.54 \cdot 10^{-5}$	$4.23 \cdot 10^{-8}$
$b_f(\theta)$	$-4.29 \cdot 10^{-3}$	$4.74 \cdot 10^{-5}$	$-1.21 \cdot 10^{-7}$	$1.17 \cdot 10^{-10}$	$-3.92 \cdot 10^{-14}$
Bainite					
$a_b(\theta)$	$-2.33 \cdot 10^{1}$	$6.91 \cdot 10^{-2}$	$1.36 \cdot 10^{-5}$	$-1.63 \cdot 10^{-7}$	$1.07 \cdot 10^{-10}$
$b_b(\theta)$	$-8.31 \cdot 10^{-1}$	$4.98 \cdot 10^{-3}$	$-1.11 \cdot 10^{-5}$	$1.09 \cdot 10^{-8}$	$-4.03 \cdot 10^{-12}$

Transformation Parameters of Austenite During Heating				
a_a^a	b_a^a	a_a^b	b_a^b	c_a^b
0.30392	$1.662 \cdot 10^{-4}$	$2.25 \cdot 10^{13}$	8.12	$2.634 \cdot 10^{-4}$

Transformation Parameters of the Diffusionless Martensitic Phase Transformation	
"S"-Shape	$a = 99.0 \quad b = 2.87 \cdot 10^{-2} \quad c = 613.15$
"S"-Shape	$a_{opt} = 40.79 \quad b_{opt} = 0.030346 \quad c_{opt} = 640.17$
KM	$\gamma = -0.01354$

Transformation Temperatures and Latent Heat				
	Austenite	Ferrite	Bainite	Martensite
Θ_α^s [K]	1003.15	1023.15	873.15	613.15
L_α $\left[10^7 \frac{\text{J}}{\text{m}^3}\right]$	-65.2	65.2	65.2	32.6

linear dependence of the elastic behavior on temperature, Poisson's ratio has been set to be constant for all phases (see Table 7.5). The thermal expansion coefficients of the phases were obtained from dilatation tests published by Bok et al. (2015). The specific heat capacity and the heat conductivities of the phases have been taken from Tang et al. (2014). In accordance with the derived model, the heat capacities have been set to be constant (see Table 7.5). However, Cheng et al. (2014) and Tang et al. (2014) found a temperature dependence of the specific heat capacities in experimental investigations (see Figure E.6). This temperature dependence is neglected in the simulations. The heat conductivities of the phases are modeled to be linear, whereby the fitting is done by a least square approach (cf. equation (7.2)).

In recent research, for the steel 22MnB5, the yield curves of the different phases, which can occur in the hot stamping process, have been investigated experimentally by several authors, e.g. Hochholdinger (2012) and Bok et al. (2015). They found that the ferritic, bainitic, and martensitic phases are approximately loading rate independent. However, even

Table 7.5: Young's moduli (Miokovic, 2005; Schwenk et al., 2012), Poisson's ratios, thermal expansion coefficients (Bok et al., 2015), transformation strains, specific heat capacities, heat conductivities (Tang et al., 2014), and TRIP parameters of phases of 22MnB5

	Austenite	Ferrite/Pearlite	Bainite	Martensite
E_α: m^E_α $\left[\frac{10^8\,\mathrm{Pa}}{\mathrm{K}}\right]$	-1.63	-1.0	-0.8276	-0.73
n^E_α $\left[10^{11}\,\mathrm{Pa}\right]$	2.55	2.51	2.41	2.36
ν_α	0.3	0.3	0.3	0.3
α_α $\left[\frac{10^{-5}}{\mathrm{K}}\right]$	2.05	1.55	1.38	1.38
ζ_α $\left[10^{-3}\right]$	-2.5	2.5	2.5	1.12
κ_α $\left[\frac{\mathrm{J}}{\mathrm{kgK}}\right]$	748.74	594.91	570.51	570.51
Λ_α: m^Λ_α $\left[\frac{10^{-3}\,\mathrm{W}}{\mathrm{mK}^2}\right]$	14.838	34.393	1.7991	1.7991
n^Λ_α $\left[\frac{\mathrm{W}}{\mathrm{mK}}\right]$	7.2152	59.482	40.521	40.521
$k^l_{0,\alpha}$ $\left[\frac{10^{-11}}{\mathrm{Pa}}\right]$	-	7.95	2.522	5.608
$k^l_{1,\alpha}$ $\left[\frac{10^{-18}}{\mathrm{Pa}^2}\right]$	-	1.576	1.2176	2.4778
$k^w_{0,\alpha}$ $\left[\frac{10^{-11}}{\mathrm{Pa}}\right]$	-	17.17	8.887	5.7199
$k^w_{1,\alpha}$ $\left[\frac{10^{-11}}{\mathrm{Pa}}\right]$	-	6.5156	7.909	3.0533
$k^w_{2,\alpha}$ $\left[\frac{10^{-3}}{\mathrm{Pa}^{k^w_{3,\alpha}}}\right]$	-	0.0	0.345	2.4057
$k^w_{3,\alpha}$ $[-]$	-	7.24	2.5395	1.5268

at room temperature, the austenitic phase shows viscoplastic effects (Figure E.1). In this work, viscoplastic effects are neglected and, since the forming step in hot stamping is performed with high stamp velocities, the parameters of the Swift hardening potential (4.47) are fitted to the strain-stress curves gained by the fastest forming (see left-hand side of Figure 7.5). For the modeling of the hardening of ferrite, bainite, and martensite, the temperature-dependent Swift hardening potential (4.47) is also used. The corresponding parameters are listed in Table 7.6.

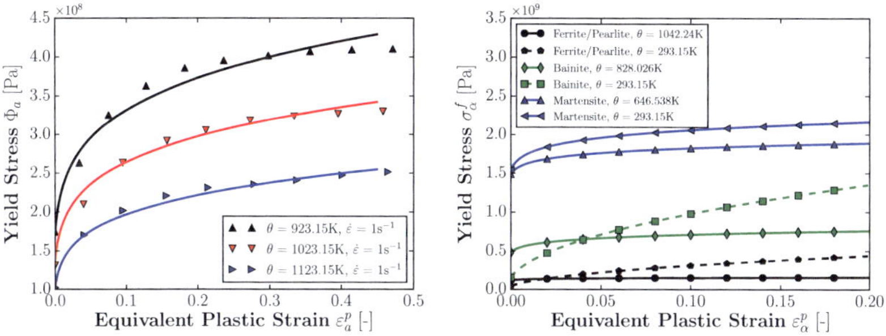

Figure 7.5: Yield curves of austenite (left) (Hochholdinger, 2012) and ferrite, bainite, and martensite (right) (Bok et al., 2015) of 22MnB5

Table 7.6: Swift parameter of austenite and ferrite, bainite, and martensite (Bok et al., 2015) of 22MnB5

	Austenite	Ferrite/Pearlite	Bainite	Martensite
$s_\alpha^{f,0}$ $[10^6\text{Pa}]$	1418.45	1022.0	2958.0	2449.0
$s_\alpha^{f,1}$ $\left[10^6 \frac{\text{Pa}}{\text{K}}\right]$	−1.0026	−1.1033	−3.7276	−1.0667
$s_\alpha^{e,0}$ $[-]$	0.17446	0.5247	0.4871	0.0742
$s_\alpha^{e,1}$ $\left[\frac{10^{-4}}{\text{K}}\right]$	0.0	−6.0	−7.0	−0.6
ε_α^0 $[10^{-3}]$	2.59237	1.8	2.0	2.0

7.2.2 Phase Transformation Concerning Parameters

The transformation start temperatures Θ_α^s for austenite in the heating case, and ferrite/pearlite, bainite, and martensite in the cooling case are determined by the TTT diagram published by He et al. (2010) and the dilatation tests determined by Bok et al. (2015). They are given in Table 7.4.

The phase transformation accompanying transformation strain is determined by the same dilatation tests. Similar to the steel 42CrMo4, the transformation strains are related to the respective transformation finish temperatures of the arising phases (see Figure 7.3). The values of the mean relative volume change during phase transformation are obtained by use of the optimization method MIDACO (see Schlueter et al., 2012). For the transformation strain from ferrite/pearlite to austenite, and from austenite to martensite, one obtains $\zeta_a = -0.0025$, and $\zeta_m = 0.00112$, respectively. The transformation strain of ferrite and bainite is set to be $\zeta_b = 0.0025$.

The latent heats, which are released or consumed during phase transformation are taken from Bok et al. (2015) and are listed in Table 7.7.

Diffusion-Driven Phase Transformation

Unlike the method used by Hömberg (1996), which works with an interpolation of the JMAK parameters gained from the TTT curve by determination of the transformation start and end temperature, Bok et al. (2015) found the transformation parameters for his model by adjusting the resulting volume fractions to the hardness measured in thermomechanical experiments. Thereby, the overall hardness is given by the volume average of the phases' hardness and the phase hardness as a function of the cooling rate. In this work, the parameters of the JMAK and the KV model are fitted that way, that the transformation start curve and end curve of the experimental TTT curve and the numerical TTT curve coincide. For this purpose the least square fit between the experimental and numerical curves are done using the optimization tool MIDACO. The resulting TTT diagram is depicted in Figure 7.6, where the experimentally determined TTT curves published in Guo (2011), Tang et al. (2014), and He et al. (2010) are shown. The latter has been used for parameter identification. Since the transformation start and end isolines computed with the JMAK and the KV model are in good agreement with the experimental results, the difference of the models is shown in Figure 7.7. In this volume fraction-temperature-time diagram, the different transformation kinetics can be considered. In the isothermal case, the transformation rate in the KV is higher at the beginning of the transformation as in the JMAK model, but decreases faster as it is the case in the JMAK model. Since no experimental information on the

Table 7.7: Phase transformation parameters, transformation temperatures, and latent heats Bok et al. (2015) of 22MnB5 (due to brevity, the units of most parameters are omitted; since SI units are used they can be easily determined from the corresponding equations)

Transformation Parameters					
JMAK-Model			KV-Model		
	Ferrite	Bainite		Ferrite	Bainite
a_α	3.264	4.169	m_α	2.752	1.329
$b_{1,\alpha}$	1.5889	5.1727	S	8.0	8.0
$b_{2,\alpha}$	9.0851	7.0755	A_α	$2.037 \cdot 10^8$	$3.637 \cdot 10^9$
$b_{3,\alpha}$	3.549	3.317	d_α	$7.81 \cdot 10^4$	$6.92 \cdot 10^4$
$a_{0,\alpha}$	1.05	0.95	n_α	0.757	0.729
$b_{0,\alpha}$	15	6.0	$r_{0,\alpha}$	3.2	2.0
$b_{1,\alpha}$	-27.2	-50.0	$r_{1,\alpha}$	-27.2	-50.0
$b_{2,\alpha}$	-17.0	-35.0	$r_{2,\alpha}$	-20.0	-30.0

Transformation Parameters of Austenite During Heating				
a_a^a	b_a^a	a_a^b	b_a^b	c_a^b
0.30392	$1.662 \cdot 10^{-4}$	$2.25 \cdot 10^{13}$	8.12	$2.634 \cdot 10^{-4}$

Transformation Parameters of the Diffusionless Martensitic Phase Transformation				
"S"-Shape	$\theta_m = 700.0$	$a = 99.0$	$b = 7.659 \cdot 10^{-2}$	$c = 700.0$
"S"-Shape	$\theta_m = 720.0$	$a_{opt} = 41.96$	$b_{opt} = 0.030017$	$c_{opt} = 669.18$
KM	$\theta_m = 668.0$	$\gamma = -0.029634$		

Transformation Temperatures and Latent Heat				
	Austenite	Ferrite	Bainite	Martensite
Θ_α^s [K]	1003.15	1042.24	828.03	see above
L_α $\left[10^3 \frac{\mathrm{J}}{\mathrm{kg}}\right]$	-89.4	89.4	66.5	81.5

transformation kinetics is available in the TTT nor CCT diagrams, no model can be preferred.

The resulting CCT curve from the TTT fitted parameters is depicted on the right-hand side of Figure 7.6. In the Figure, experimental results from Hochholdinger (2012), Naderi (2007), and MBW 1500 (2016) are compared with the numerical results. Since a great difference between experiment and model can be observed, the JMAK and the KV model have been extended to take the cooling rate into account. The result is shown on the right-hand side of Figure 8.2 with the associated parameters given in Table 7.7. Note, that due to the nature of the extensions, the isothermal behavior of the JMAK and KV model is untouched.

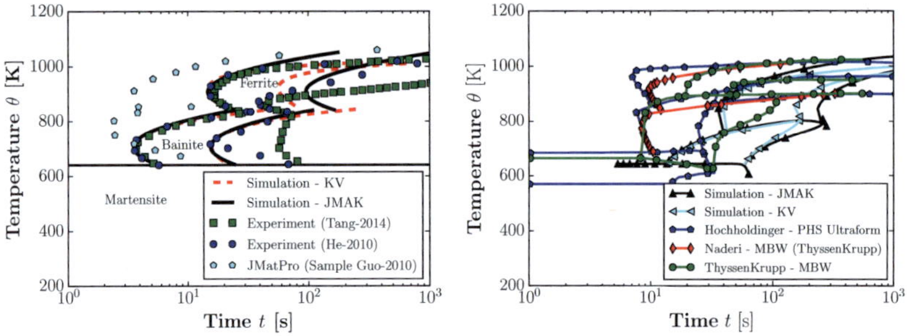

Figure 7.6: Comparison of the experimentally and numerically determined TTT (left) and CCT curves (right) of 22MnB5

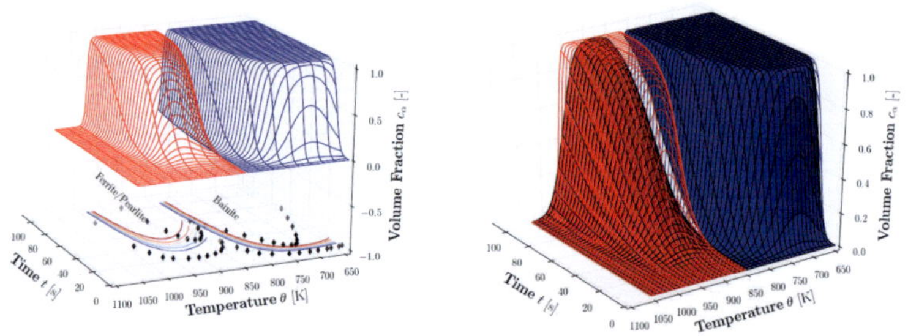

Figure 7.7: Comparison of the TTT curves determined with the KV model and experimental results (left) and comparison of the TTT curves determined with the JMAK (surface) and the KV model (wireframe) (right)

For the description of the austenitization, the JMAK model is used. The corresponding parameters have been determined by using the heating part of dilatation tests. The resulting parameters are given in Table 7.7.

Diffusionless Phase Transformation

Similarly to the steel grade 42CrMo4, the parameters of the models describing the diffusion-driven phase transformation are identified. The parameter γ in the Koistinen-Marburger model (4.107) is identified by assuming a purely austenitic-martensitic phase transformation leading to $\gamma = \ln(1 - c_{mf})/(M_s - M_f)$. With the martensite formation finish temperature $M_f = 570.15$K for the steel 22MnB5 and $c_m(\theta = M_f) = 0.99$, the transformation parameter reads $\gamma = -0.02963\text{K}^{-1}$. A similar approach for the "S"-shape model (4.108) with the relations (7.6) leads to the parameters for the "S"-shape model $a = 99.0$, $b = 0.07659\text{K}^{-1}$, and $c = 700$K, respectively (cf. Table 7.7). However, the fitting of the numerically-determined dilatation curve to the experimental one results in the optimized parameters for the "S"-shape model $a_{\text{opt}} = 41.96$, $b_{\text{opt}} = 0.030\text{K}^{-1}$, and $c_{\text{opt}} = 669.18$K. Subsequently, this optimized parameters are used in further numerical considerations.

7.2.3 The TRIP Parameters of 22MnB5

The TRIP parameters k_α^{tr} are determined by considering dilation tests performed under external load. Bok et al. (2014), Cheng et al. (2014), and Tang et al. (2014) published such tests with different tensile stress regimes. The latter are used to identify the TRIP parameters of the

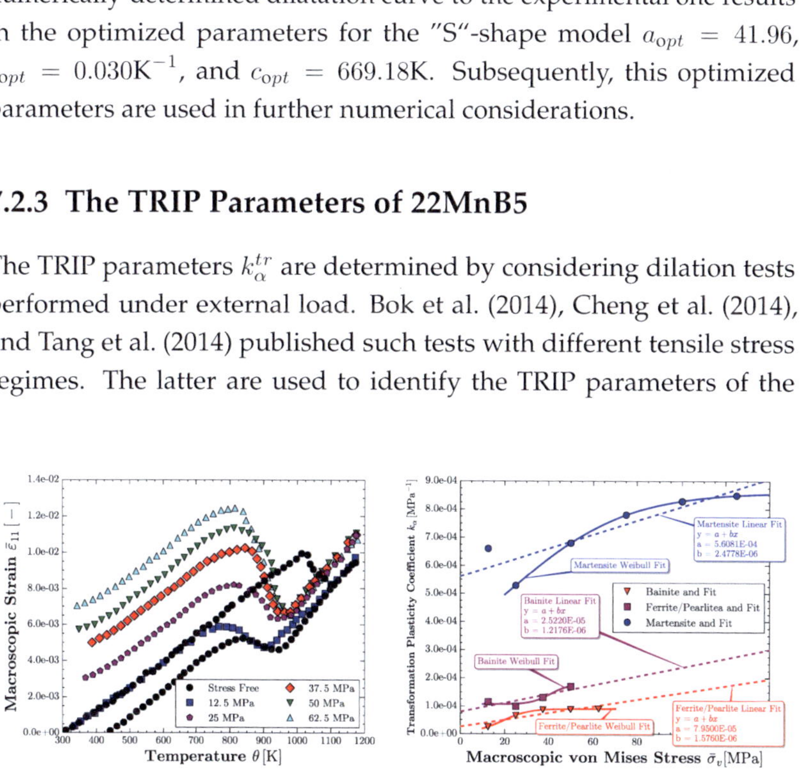

Figure 7.8: Dilatation tests of 22MnB5 under external tensile load (left) and TRIP parameters of the phases ferrite, bainite, and martensite dependent on the applied tensile stress (dots: experiment, solid lines: Weibull fit, dashed lines: linear fit) (right)

modified Leblond's equation (5.29). In such dilatation tests, a tensile stress is usually applied after the austenitization of the specimen and the strain in tensile direction is measured during quenching (see left-hand side of Figure 7.8). For identification, it is assumed that the corresponding phase transformation is pure, i.e. if only ferrite, bainite, or martensite occur during phase transformation in the dilatation test. Furthermore, it is assumed that the applied stress is constant during the entire process. With these assumptions, after an integration of the TRIP evolution equation (5.29), the TRIP strain reads

$$\varepsilon_a^{tr} = \frac{3}{2} k_\alpha^{tr} \varphi_\alpha(c_\alpha) \bar{\boldsymbol{\sigma}}'. \tag{7.7}$$

With relation (5.30), the macroscopic TRIP strain reads

$$\bar{\boldsymbol{\varepsilon}}^{tr} = \frac{3}{2} k_\alpha^{tr} \bar{\boldsymbol{\sigma}}' \tag{7.8}$$

if the entire parent phase is transformed. For the uniaxial tension, the TRIP parameter is dependent on the applied macroscopic tensile stress and the resulting TRIP strain in tensile direction (see, e.g., Tang et al., 2014)

$$k_\alpha^{tr} = \frac{\sigma_{11}}{\bar{\varepsilon}_{11}^{tr}}. \tag{7.9}$$

The TRIP parameters, which are identified by this method, are depicted on the right-hand side of Figure 7.8. For the application in the macroscopic phenomenological model and TMM model, a constant, a linear (5.31), and a Weibull fit (5.32) are considered. The associated parameters of the linear and Weibull fit are listed in Table 7.5. For the constant fit, the constant part of the linear fit is used

$$k_\alpha^{tr} = k_{0,\alpha}^{l}. \tag{7.10}$$

7.3 Influence of the Temperature-Dependent Elasticity and Plasticity on the Specific Heat Capacity

The influence of the temperature-dependent elastic parameters and the temperature-dependent Voce parameters on the specific heat capacity is depicted in the top and bottom of Figure 7.9 for 42CrMo4 and 22MnB5, respectively. Since the maximum relative deviation of

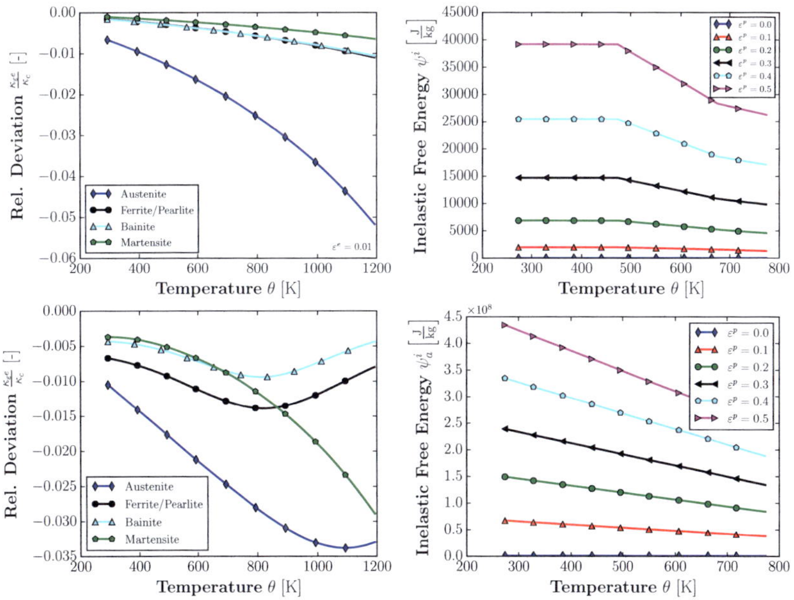

Figure 7.9: Relative deviation of the thermoelastic contribution κ_{ε^e} for a uniaxial strain of $\varepsilon^e = 0.01$ (top left), and the inelastic part of the Helmholtz free energy of austenite for different equivalent plastic strains (right) of 42CrMo4 (top) and 22MnB5 (bottom)

the part $\kappa_{\varepsilon^e} = 2(\theta/\rho)\boldsymbol{\varepsilon}^e \cdot \partial\mathbb{C}/\partial\theta\,[\boldsymbol{\alpha}]$ from the constant part κ_c is about 6% for austenite for 42CrMo4 and 3.5% for 22MnB5, the thermoelastic part of the specific heat capacity can be ignored for both steel grades. Furthermore, the inelastic part of the Helmholtz free energy is piece-wise linear in temperature for 42CrMo4. Except at the transition temperatures, the linearity leads to a vanishing second derivative of the Voce hardening potential $\kappa_{\psi^i} = \theta \partial^2 \psi^i / \partial \theta^2$. Thus, this influence on the specific heat capacity can be neglected and the specific heat capacity is approximately constant for 42CrMo4. Similarly, the Swift hardening potential of 22MnB5 of the austenitic phase, for example, shows a linear behavior (see bottom right-hand side of Figure 7.9) leading to a vanishing specific heat contribution due to inelastic effects $\kappa_{\psi^i} = \theta \partial^2 \psi^i / \partial \theta^2$. Therefore, the assumption of an approximately constant specific heat capacity of the phases is in good agreement with the observations.

Chapter 8

The Simulation of the Process Chain Hot Stamping[1]

8.1 Validation of the Models

The validation of the TMM model is performed for both steel grades 42CrMo4 and 22MnB5. Thereby, the phase transformation behavior of 42CrMo4, which is identified by using only the TTT diagram, is validated by considering the transformation behavior in the continuous cooling case, i.e. the CCT diagram. For the steel 22MnB5, the TTT diagram was used to optimize the JMAK and KV parameters. Furthermore, it was observed, that the phase transformation behavior in the continuous cooling case deviates from the measured CCT curves. This led to the rate extension of the models resulting in better agreement with the experimental considerations.

In the non-isothermal case, there are two methods to implement the JMAK equation by using Scheil's additivity rule. Both methods are compared, the isothermal step method is validated, and the maximum time step to get reasonable results is considered.

[1] The validation of the TMM model with respect to the steel grade 42CrMo4 and the results of the investigation of the influence of the transformation strain and latent heat are shwon in the paper "Hashine-Shtrikman type mean field model for the two-scale simulation of the thermomechanical processing of steel" (Neumann and Böhlke, 2016). For the sake of completeness the results are represented.

The thermomechanical coupling of the TMM model and the macroscopic phenomenological reference model are validated by dilation tests with and without external load. Most notably, the dilatation behavior in the quenching part shows a deviating curve shape for 22MnB5 and 42CrMo4. While the steel grade 42CrMo4 shows an "S"-shape transition behavior from austenite to martensite, the steel transition from austenite to martensite is exponential for 22MnB5. It is found, that the impact of the homogenization method is dependent on the type of implementation of the TMM model. While Neumann and Böhlke (2016) used a direct implementation of the TMM model for the steel 42CrMo4, resulting in a minor role of the homogenization scheme on the dilatation behavior, the incremental implementation leads to a greater influence of the homogenization scheme on the linear thermomechanical behavior.

8.1.1 The Isothermal Step Method of the JMAK Model

The validity and the time step size dependence of the isothermal step method of the diffusion-driven phase transformation are investigated using the transformation of austenite into ferrite, which serves as an example. To show the validity of the implementation, the resulting volume fraction of ferrite/pearlite is compared with the numerical solution of the rate law derived directly from Scheil's rule (4.100). Hömberg (1995) showed the existence and the uniqueness of the solution of equation (4.100) and Hömberg (1996) applied an explicit method to simulate the Jominy End-Quench test of an eutectoid carbon steel. The solution received from an implicit integration of the rate law, derived directly from Scheil's rule, is used as a reference for the isothermal step method. The implicit discretization in time is given in equation (6.6).

The initial austenitic microstructure at the initial temperature $\theta_0 = 1173.15\text{K}$ is cooled linearly $\theta(t) = \theta_0 + \zeta t$ with a cooling rate $\zeta = -0.1\text{K/s}$, to get a purely ferritic microstructure. Using a constant time discretiza-

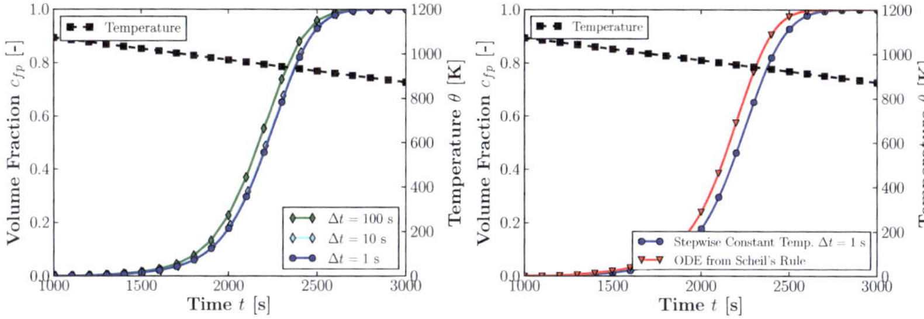

Figure 8.1: Comparison between different time discretizations of the JMAK implementation (left) and comparison between the JMAK implementation and the implicit time integration of equation (4.100) (right)

tion of $\Delta t = 0.1$s and the initial volume fraction $c_{fp} = 10^{-10}$, on the right-hand side of Figure 8.1, the numerical solution of equation (4.100) for the steel 42CrMo4 is depicted. The slight deviation of the different methods can be accepted with regard to the lower complexity of the isothermal step method.

The time step size dependence of the isothermal time step method is depicted on the left-hand side of Figure 8.1. For the linear cooling, a constant time step size of $\Delta t = 10$s and lower is sufficient to obtain a good approximation of the volume fraction evolution.

8.1.2 The Phase Transformation Behavior

The models describing the diffusion-driven phase transformation behavior of both the steel grade 42CrMo4 and 22MnB5 are validated for the non-isothermal case. The comparison between the simulation and the experiments for the continuous cooling process is depicted in Figure 8.2. For the steel grade 42CrMo4, the prediction of the start of the transition of austenite into ferrite/pearlite agrees well with experimental results for slow cooling rates (0.1K/s up to 4K/s). Using the JMAK model, the critical cooling rate to get a microstructure without ferrite is overestimated by about 3.5s. In contrast, the critical cooling rate to get a pure martensitic microstructure is underestimated by 40K/s. The start of the transformation from austenite to bainite happens between 3s and 7s later in the simulation when compared with the experiments.

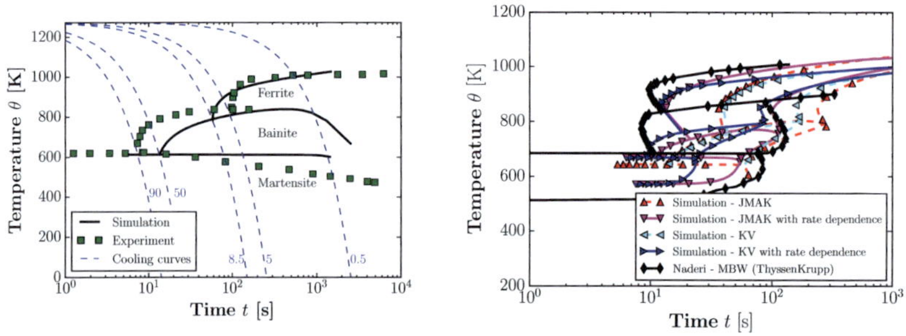

Figure 8.2: Experimentally- and numerically-determined CCT diagrams of 42CrMo4 (left) and 22MnB5 (right)

For the steel grade 22MnB5, due to the rate extension of the JMAK and KV model, the critical cooling rates to get a microstructure without ferrite and to get a pure martensitic microstructure agree well with the experimental results found by Hochholdinger (2012), Naderi (2007), and MBW 1500 (2016). However, the transformation behavior of both the JMAK and the KV model appear to have a longer incubation time and shorter transformation time compared to the experiments.

The deviation between the results of the simulation and the experiments in the bainitic transformation can be explained by the ansatz to model the highly complex transformation behavior of bainite by taking only the diffusion-driven phenomena into account. A reason for the differences between the simulation and the experiment in the ferritic transformation could be the fact that the two different phases ferrite and pearlite are assumed to be one phase. To improve the prediction of the transformation kinetics of the steels and to capture the complex transformation behavior of the diffusion-driven transformation, an extended version of the Scheil's rule can be used as proposed in Réti and Felde (1999).

8.1.3 The Thermomechanical Coupling of the TMM and Macroscopic Reference Model

The thermomechanical coupling is validated by using experimental results from dilatation tests. A ferritic steel is heated since a pure austenitic microstructure is obtained. Thereafter, the austenitized steel is quenched to room temperature. The temperature path of the exper-

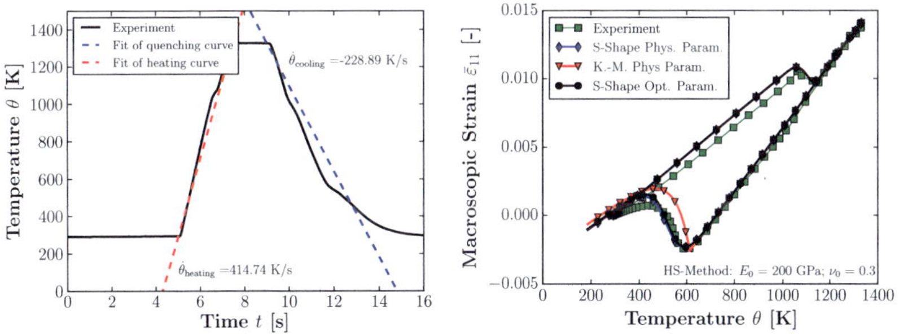

Figure 8.3: The imprinted temperature load (left) for the dilatation test and the comparison of different methods and corresponding parameters (right) of 42CrMo4

imental procedure is shown on the left-hand side of Figure 8.3 for the short time austenitization of 42CrMo4 and 22MnB5. On the right-hand side of Figure 8.3 for 42CrMo4, the comparison between the experiment and model gained by both the KM model and the "S"-shape model are depicted with different parameters each. For 42CrMo4, the "S"-shape approach leads to a good agreement between simulation and measurement. However, the transformation behavior from austenite to martensite of 22MnB5 shows an exponential behavior in the dilatation measurement (see Figure 8.5). In this case, the KM model leads to a better agreement between simulation and measurement.

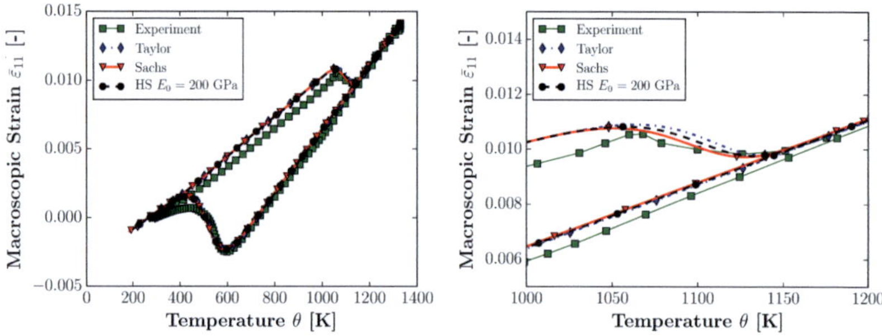

Figure 8.4: Comparison of homogenization methods in the dilatation test (left) and detailed (right) for 42CrMo4

The influence of the chosen homogenization method on the macroscopic behavior during the dilatation test is depicted in Figure 8.4 for 42CrMo4 and 8.5 for 22MnB5. Here, the Sachs, Taylor, and Hashin-Shtrikman solutions are compared. The impact of the homogenization scheme on the final thermomechanical behavior in the dilatation test is dependent on the kind of implementation of the TMM model. Neumann and Böhlke (2016) used a direct implementation of the TMM model leading to the results in Figure 8.3, while, in this work, an incremental implementation (cf. Chapter 6) is chosen to be able to simulate the finite deformation process via the ULF. However, this kind of implementation leads to a greater dependence of the effective thermomechanical behavior on the homogenization scheme (see upper left-hand side of Figure 8.5).

To validate the TRIP behavior of the TMM model, the macroscopic phenomenological model (see Section 5.1) is used as a reference. For this purpose, the dilatation test is modified. After heating, the high temperature level is held for 4s and a tensile load is applied. Thereafter, the tensile loaded specimen is quenched. Thereby, the tensile stress is

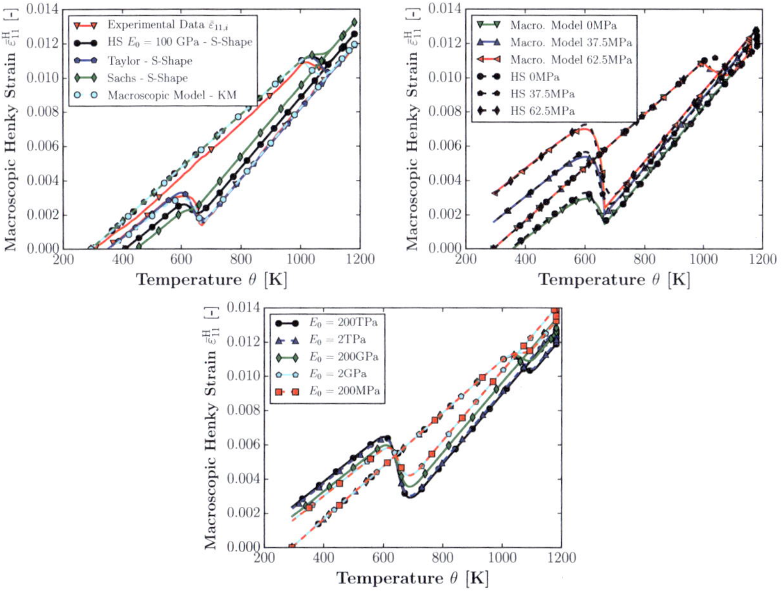

Figure 8.5: Comparison of homogenization methods in the dilatation test (upper left) and the TRIP effect in the dilatation test: comparison of models (upper right) and comparison of homogenization estimate (bottom) for 22MnB5

kept constant throughout the entire quenching step. The resulting curves are depicted on the upper right-hand side of Figure 8.5, where for the macroscopic reference model the KM model and for the TMM model, the "S"-shape model has been used with optimized parameters for both transformation models (see Table 7.7). In these simulations, the homogenization parameters, i.e. Young's modulus and the Poisson's ratio of the isotropic homogeneous comparison material, read $E_0 = 200\text{TPa}$ and $\nu_0 = 0.3$, respectively. Despite the different transformation behavior, the models are in good agreement with a maximum relative deviation $d_{L^2}^{rel}$ of 2.57% of the cooling-associated curves.

Furthermore, the influence of the homogenization estimate is depicted on the bottom of Figure 8.5. In the regime between $E_0 = 200\text{TPa}$ and

$E_0 = 200$MPa with a Poisson's ratio of $\nu_0 = 0.3$, the effective thermo-mechanical behavior changes with a maximum relative deviation of the final state of about 55%. A Young's modulus of the homogeneous comparison material beyond this regime leads to minor changes of the effective behavior.

Note, that since the incremental method for the implementation of the models is used in the context of the ULF, the Hencky strain is used. For small deformations, the Hencky strain corresponds to the infinitesimal strain.

Due to the fact, that the previous dilatation tests were also used for the optimization of the diffusionless phase transformation parameters, another dilatation test with a more complex cooling curve is considered (see left-hand side of Figure 8.6). After austenitization of the 42CrMo4 steel, the specimen is cooled in three steps: quenching until a temperature level of 623.15K is reached, leaving the temperature constant at this level for 50s to initiate the bainitic transformation, and further quenching until room temperature is reached to transform the retained austenite into martensite. By this procedure, a microstructure consisting of bainite and martensite in the specimen is obtained. On the right-hand side of Figure 8.6, the experimental dilatation curve is compared with the numerical ones gained by use of the Taylor, Sachs, and Hashin-Shtrikman method.

The comparisons of experimental dilatation curves with numerical results reveal two nuisances: first, the transformation model from ferrite to austenite optimized for the short time austenitization shows a deviation from the experiments in the long time austenitization and second, the transformation strain in the combined transformation from austenite to bainite and martensite is overestimated.

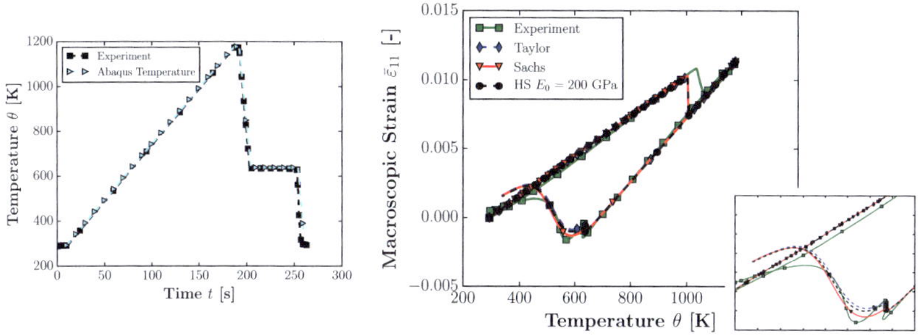

Figure 8.6: Temperature path (left), comparison of results gained by use of different homogenization methods with experimental results and detailed depiction (right) for 42CrMo4

8.2 The Influence of the Transformation Strain and the Latent Heat on the Final State of the Thermomechanically Treated Sheet Metal

As can be seen in Figure 8.4 and 8.6, the regime is tight, which is spanned by the results with the first-order bounds methods, i.e. Sachs and Taylor, in a dilatation test for the 42CrMo4 steel grade. Since only the linear thermal strain and the constant transformation strain are considered in a dilatation test, either of the classical approaches, i.e. the first-order bounds, yield a good approximation of the process. On the other hand, the incremental implementation, reveals the high influence of the homogenization scheme on the final thermomechanical behavior even in the linear case.

The aim of this section is to show, that, independently of the implementation method, the classic approaches span a wide regime of possible stress or shape responses for processes beyond linear thermal or constant transformation strain. It is shown, that in the case of a thermo-elasto-plastic process, the homogenization estimate plays an important role. Furthermore, the impact of the latent heat and transformation strain on the final stress and shape state is investigated.

For this purpose, the developed and validated model with a direct implementation as described in Neumann and Böhlke (2016) is applied to a thermomechanical boundary initial value problem. A sheet of 42CrMo4 steel with a hole in it is thermomechanically treated. The initial state is austenitic at a temperature of $\theta_0 = 1173.15\mathrm{K}$. While the sheet is quenched by a heat transfer to a water polymer mixture, a mechanical load in terms of a displacement in y-direction, i.e. the load describing vector $(u, v, w)^{\mathrm{T}} = (0, 0.009, 0)^{\mathrm{T}}\mathrm{m}$, is applied at the front of the sheet metal. The geometry and the boundary conditions of the boundary initial value problem are depicted in Figure 8.7 (upper left). The heat transfer coefficient between the steel 42CrMo4 and the water polymer mixture used for quenching are dependent on temperature. This dependence is given in Schwenk et al. (2012) and is realized in the subroutine FILM by ABAQUS/Standard.

Using this complex thermomechanical problem, the influence of the following parameters on the

- Deformation in z-direction of the points along Path A and
- Stress, strain, temperature, and martensite evolution in the elements A, B, C

are investigated:

- Homogenization method, that are the first-order bound of Taylor and the estimate of Hashin-Shtrikman-type with different stiffnesses of the homogeneous comparison material $\mathbb{C}_0$,
- Latent heat released during phase transformation, and
- Transformation strain occurring during phase transformation.

Considering Figure 8.7, the Taylor solution and the influence of the stiffness of the comparison material on the macroscopic stress $\bar{\sigma}_{22}$ in A, the macroscopic strain $\bar{\varepsilon}_{22}$ in A, and the final deformation of Path A in z-direction w along the path coordinate y_{path} are depicted. The magnitude of the isotropic stiffness of the comparison material is represented by

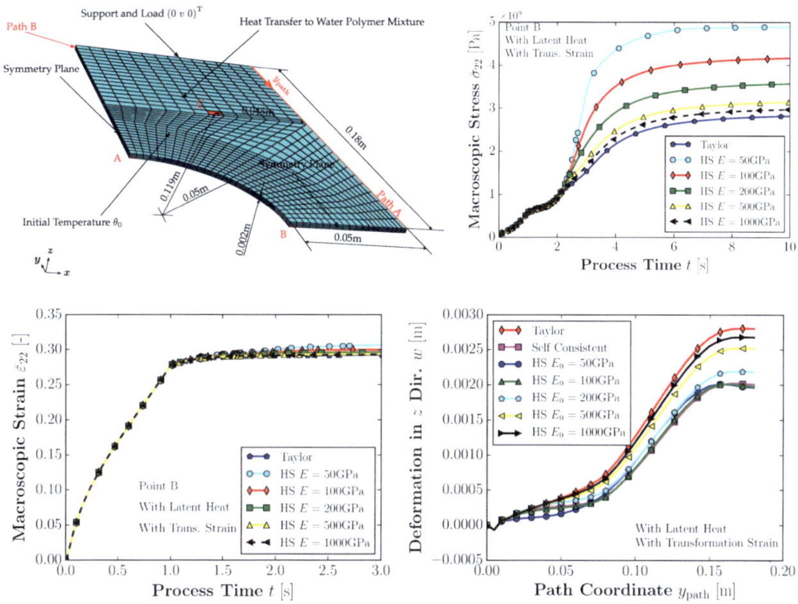

Figure 8.7: Geometry of the sheet metal with a hole in int with boundary conditions (top left), influence of $\mathbb{C}_0$ on the stress development (top right), strain development (bottom left) at Point B, and the deformation along Path A at the end of the process (bottom left)

the Young's modulus E_0 and a constant Poisson's ratio $\nu_0 = 0.3$. Using the Frobenius norm $\|\mathbb{C}_0\|_F = \sqrt{\mathbb{C}_0 \cdot \mathbb{C}_0}$, the magnitude of the isotropic stiffness of the comparison material is a linear function of E_0

$$\|\mathbb{C}_0\|_F = \sqrt{\frac{1}{(1-2\nu_0)^2} + \frac{5}{(1+\nu_0)^2}}\, E_0 \overset{\nu_0=0.3}{\approx} 3.035 E_0. \tag{8.1}$$

The lower first-order bound of Sachs causes convergence problems for this specific initial boundary value problem. However, the figure clearly displays that the stress development and the final shape are highly influenced by the stiffness of the comparison material $\mathbb{C}_0$. Therefore, there is a wide range for the actual material behavior between the first-order bounds. At this point, the potential of the considered model becomes evident. By varying the magnitude of the stiffness of the comparison material $\|\mathbb{C}_0\|_F$, the first-order bounds, as well as each material behavior between these bounds can usually be adjusted.

Note, in a simple tensile or shear test at Gauss point level, the energetic bound by Taylor is the upper bound and the bound by Sachs the lower bound in the stress-strain diagram. In the macroscopic stress-time diagram depicted in Figure 8.7, the bound by Taylor occurs as the lower bound. When combining the macroscopic stress-time diagram and the macroscopic strain-time diagram with a macroscopic stress macroscopic strain diagram, the Taylor bound appears as the upper bound again.

Figure 8.8 shows the influence of the latent heat and the transformation strain on the deformation in the z-direction along path y_{path}, the evolution of temperature, the martensitic volume fraction, and the macroscopic stress and strain. The computations were done using the Hashin-Shtrikman estimate with Young's modulus of the comparison material of $E_0 = 200$GPa and a Poisson's ration of $\nu_0 = 0.3$. The deviations of the resulting curves are measured with the L^2-norm-induced metrics (2.2) and the relative deviation (2.3).

In Table 8.1, the relative deviations of the resulting curves, taking the latent heat into account and neglecting the latent heat, and of the resulting curves, taking the transformation strain into account and neglecting the transformation strain, are shown. Thereby, the following abbreviations are introduced: lh taking latent heat into account, $nolh$ neglecting latent heat, ts taking transformation strains into account, and $nots$ neglecting transformation strains. Comparing the deviations of the macroscopic strains and stresses in Point B, and deformation in the z-direction of Path A given in percent in Table 8.1, one observes that the impact of the latent heat and transformation strain on the overall strain is of little importance. Both influences are more important for the resulting stress and the final deformation of the body.

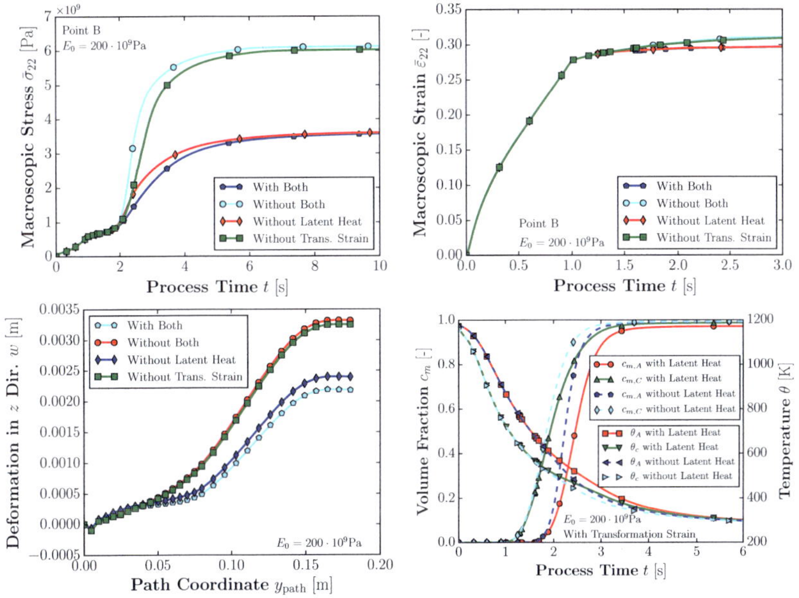

Figure 8.8: Influence of latent heat and transformation strain on the stress development (top left), strain development (top right) at point B, the deformation along Path A at the end of the process (bottom left), the martensitic volume fraction, and the temperature (bottom right)

While the influence of the transformation strain is slightly reduced with occurring latent heat in the strain and stress development, the influence of the transformation strain on the final shape given by the deformation in the z-direction of Path A is clearly increased if latent heat is taken into account. For the latent heat, the same effect is considered. The influence of the latent heat on the strain and stress is reduced with occurring transformation strain and the influence on the final shape is increased with occurring transformation strain. However, considering the stress time diagram and the strain time diagram in Figure 8.8, the stress and strain reach the same state at the end of the process independently of latent heat. On the other hand, the final shape is slightly influenced by released heat during phase transformation.

Altogether, considering small influences of the latent heat on the temperature and martensite evolution depicted in Figure 8.8 (bottom right), one can conclude that the transformation strain is more important in thermomechanical simulations than the latent heat.

Table 8.1: Influence of latent heat and transformation strain on the stress and strain at point B and on the deformation along Path A at the end of the process measured by use of the L^2-norm-induced metrics

	Strain $\phi = \bar{\varepsilon}_{22}$	Stress $\phi = \bar{\sigma}_{22}$	Def. in z-Dir. $\phi = w$
$d^{\text{rel}}_{L^2}(\phi^{lh}, \phi^{nolh})\|_{ts}$	0.0768 %	3.389 %	10.59 %
$d^{\text{rel}}_{L^2}(\phi^{lh}, \phi^{nolh})\|_{nots}$	0.3111 %	4.865 %	2.421 %
$d^{\text{rel}}_{L^2}(\phi^{ts}, \phi^{nots})\|_{lh}$	3.763 %	70.73 %	53.71 %
$d^{\text{rel}}_{L^2}(\phi^{ts}, \phi^{nots})\|_{nolh}$	4.113 %	71.86 %	42.45 %

Summary and Conclusions

Considering the dilatation tests with linear thermal and the constant transformation strain, the influence of the homogenization scheme in the direct implementation appears to be negligible, but important in the incremental implementation. To show that the homogenization scheme has a huge impact on the final state of a thermomechanically treated part even in the direct implementation in a process beyond linear or constant strains, a numerical experiment is performed taking nonlinear strains into account. At this point, the merit of the suggested TMM model over the existing phenomenological models becomes clear. Furthermore, on this occasion, the influence of the latent heat and the transformation strain on the macroscopic stress, macroscopic strain, and the final shape of a sheet of 42CrMo4 with a hole in it is considered. It is found that

- The homogenization scheme captures both the upper and lower first-order bounds. With increasing magnitude of the stiffness of the comparison material, the solution approaches the Taylor bound.

- In thermoelastic processes with a direct implementation of the TMM model, the first-order bounds are close to each other for the 42CrMo4 steel. In this case, the homogenization scheme has a minor influence on the strain state. On the other hand, in an incremental implementation of the TMM model, the homogenization estimate plays an important role.
- In processes beyond linear thermal strain and constant transformation strain as the dilatation test, i.e. processes where TRIP and/or plastic strains arise, the impact of the homogenization scheme on the final state of a thermomechanically-treated part is significant. The merit of the suggested model is the possibility of adjusting the material behavior via stiffness of the homogeneous comparison material.
- The influence of the latent heat on the final state of a part is low and can be neglected in simulations aiming at the final shape or stress state. However, the transformation strain plays an important role in a thermomechanical treatment simulation of steels with complex behavior, and should be considered in any case.
- The "S"-shape model for the description of the diffusionless martensitic phase transformation leads to a good agreement with experimental results for the steel grade 42CrMo4, the KM model for the steel grade 22MnB5

8.3 The Hot Stamping Process

In this section, the models that are introduced, i.e. the phenomenological reference model and the TMM model, are applied for the simulation of the hot stamping of a 22MnB5 sheet metal. Thereby, the hot stamping of the w-shape geometry, which is described in Chapter 2, is considered. The geometry of the IBVP is depicted in Figure 8.9, which is a reduced model of the quarter of the deep drawing press shown in Figure 2.5.

The symmetry planes hold as depicted in Figure 2.5. The punch, blank holder, counter punch, and die are realized by rigid body shell surfaces with a thickness of 0.1mm. The sheet metal with a thickness of 1.8mm is defined as a solid three dimensional part.

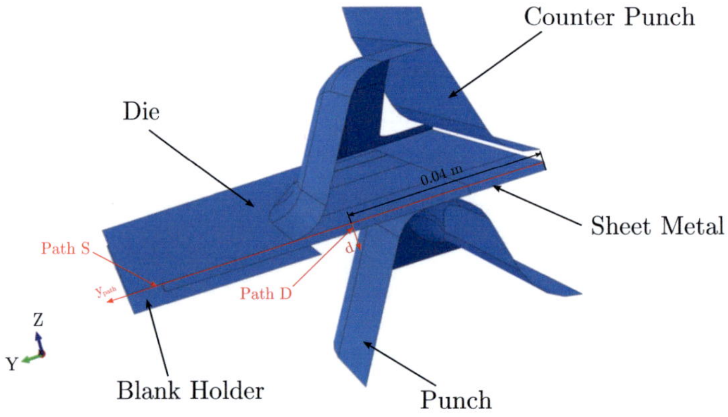

Figure 8.9: Model for the simulation of the hot stamping process of the w-shape geometry

The simulation takes place in four steps:

1. Linear heating of the sheet metal from room temperature, i.e. 293.15K, to the austenitization temperature of 1173.15K in 10s,
2. Forming of the austenitized sheet metal with a punch velocity of 30mm/s for 1s,
3. Quenching of the formed sheet metal in the closed press with a holding time of 10s, and
4. Unloading.

Thereby, the punch, blank holder, counter punch, and die are set to be isothermal with a constant temperature of 293.15K throughout the entire process. The distance between the die and the blank holder is 4.3mm and constant throughout the process. The mesh information of the associated parts are listed in Table 8.2. Since all parts except the sheet metal are rigid bodies with constant temperatures, no degrees of freedom (DOF) are present in the FE simulation for these parts. In the first step, due to the prescribed temperature and support of the sheet metal, the number of DOF is reduced even more. In the other steps, all four DOF of the sheet metal's material points, i.e., temperature and three dimensional translation, are active.

Table 8.2: Part definitions, element types in ABAQUS/Standard, element number, and degrees of freedom of the hot stamping simulation

Part	Part Definition	Element Type	Element Number	DOF
Die	Rigid Body Shell	S3T/S4RT	2413	-
Punch	Rigid Body Shell	S3T/S4RT	1595	-
Counter Punch	Rigid Body Shell	S4RT	1054	-
Blank Holder	Rigid Body Shell	S3T/S4RT	524	-
Sheet Metal	Solid	C3D8T	4591	18364

The die, blank holder, and the counter punch are fixed in space throughout the entire process. The punch is fixed in the first step, performs the forming in the second step, is fixed in the forming end position in the quenching step, and returns to the initial position in the unloading step. The contacts between the sheet metal and the die, the sheet metal and the counter punch, and the sheet metal and the punch are described by an exponential law in normal behavior with a pressure of 1GPa at a clearance of zero and zero pressure at a clearance of 10^{-5}m. As a first approximation, the tangential behavior is set to be frictionless. The thermal conductance throughout the contact surface is assumed to be pressure-dependent as found by, e.g., Abdulhay et al. (2012), Caron et al. (2014), and Merklein et al. (2009). As a first approach, the dependence is approximated linearly with a conductance of 1000W/s/m^2 at a pressure of zero and 4000W/s/m^2 at 30MPa. Other experimental and fitting models are depicted in Figure E.2. The gap dependence of the heat conductance, which is depicted in Figure E.3, is neglected in the simulations.

For the simulation, both the phenomenological model as a reference and the TMM model with different specifications each are used. According to the findings in Section 8.2, the latent heat is neglected and the transformation strain is taken into account in the simulations of the hot stamping process. The other specifications of the considered models are shown in Table 8.3.

Table 8.3: Specifications of the models considered for the simulation of the hot stamping process with the standard phase transformation specification of JMAK no temperature rate dependence and the KM model

Model Name	TRIP	Homogen.	Phase Transformation
Ref. Model Leblond-Const.	Leb. const. param.	-	standard
Ref. Model Leblond-Weibull	Leb. Weibull param.	-	standard
Ref. Model Tanaka-Weibull	Tana. Weibull. param.	-	standard
TMM SC - no TRIP	-	SC	standard
TMM SC - JMAK RD	-	SC	temp. rate dep.
TMM HS - $E_0 = 200$GPa	-	HS	standard
TMM Voigt	-	Taylor	"S"-shape
TMM SC - Leblond-Const.	Leb. const. param.	SC	standard
TMM SC - Leblond-Weibull	Leb. Weibull param.	SC	standard

The influence of the TRIP effect, the homogenization scheme, and the phase transformation model on the final shape and residual stresses in the final hot stamped part are investigated and compared with the experimental results shown in Chapter 2. The shape is considered at the cuts depicted in Figure 2.6 the residual stress at the points shown in Figure 2.8 in depth direction corresponding to Path D (cf. Figure 8.9). Furthermore, the v. Mises stress distribution along the Path S (cf. Figure 8.9) is used for comparison purposes.

On the top left-hand and right-hand side of Figure 8.10, the resulting shapes of the simulations with the reference and the TMM model are compared to the experimental findings. Both the shapes gained with the reference models and the TMM model with each specification are in

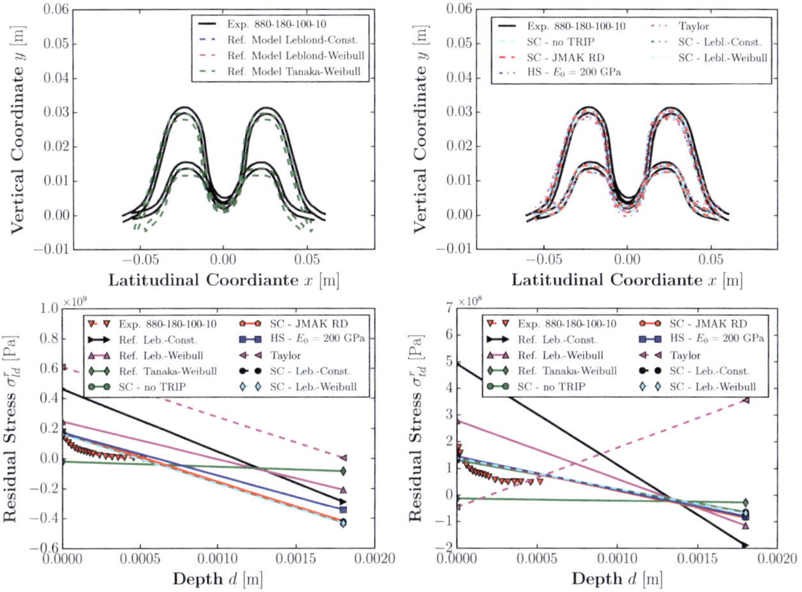

Figure 8.10: Shape of the simulated w-shaped hot stamped part at the cut A-A and B-B of the reference model and the TMM model (top left and right) and the residual stresses in longitudinal and transversal direction of the reference model and the TMM model (bottom left and right) (Neumann et al., 2017)

good agreement with the experimental results. The greatest deviation between simulation and experiment occurs in the flanks of the w-shaped part. Figure 8.11 depicts the total relative deviation (2.4) of the different models used for the simulation to the experimental result. The reference models show a deviation of about 14% and the TMM models, except the model with the Taylor approach, a deviation of about 10%. The TMM models seem to lead to better agreement with the experiment than the reference models. However, considering the u-shaped hot stamped part in the Appendix F, the reference models are in better agreement with the experiments than the TMM models. Nevertheless, both the phenomenological and the two-scale approach are quite of the same quality in view of the prediction of the final geometry of a hot stamped part. The TRIP effect and the phase transformation models have a minor influence on the final shape. The latter can be explained by the fact that in the simulations, a martensite-dominated microstructure is predicted. As depicted in Figure 8.11, which shows the spatial distribution of ferrite, bainite, and martensite, only a maximum volume fraction of about 3% and 6% of ferrite and of bainite, respectively, are computed. Due to this,

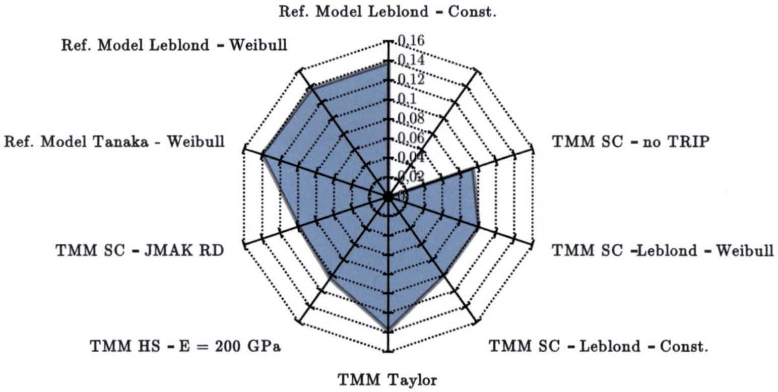

Figure 8.11: Total relative deviation of the w-shaped geometry gained from experiments and simulations

the diffusion-driven phase transformation model has a minor impact on the simulation results. On the other hand, for the simulation with the TMM Taylor model, the "S"-shape model has been used to compute the martensitic volume fraction. Since the HS scheme with Young's modulus of the comparison material of $E_0 = 200\text{GPa}$ (for the magnitude of the stiffness of the comparison material see equation (8.1)) is close to the Taylor solution, one can assume that the model for the description of the diffusionless phase transformation has a more significant influence. This finding is also considered in the residual stress analysis.

Since the deviations between the different models in view of the final shape of the hot stamped part are small, the residual stresses are called in for the valuation of the models. On the bottom left-hand and right-hand side of Figure 8.10, the residual stresses in longitudinal and transversal direction at the position which is marked in Figure 2.8 and corresponds to the Path D of Figure 8.9 are depicted. Since linear shape functions for

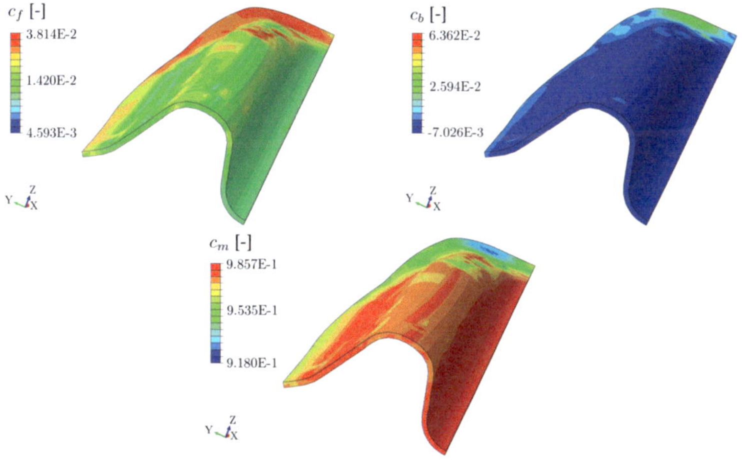

Figure 8.12: Spatial distribution of ferrite (top left), bainite (top right), and martensite (bottom) in the final hot stamped w-shaped part computed with the models using the rate independent JMAK and KM model

the FE analysis are used and the spatial discretization of the sheet metal through the thickness direction is done with one element, the residual stresses along Path D are linear.

One can observe that the reference models overestimate or underestimate the residual stresses. The predicted residual stresses of the reference model using the Tanaka approach for the TRIP effect are an almost

constant tension stress throughout the sheet metal thickness. This contradicts the experimental observations.
On the other hand, the predicted residual stresses of the TMM models are in good agreement with the experimental findings. The complex stress distribution cannot be captured by the coarse discretization used in the FE analysis, but the tendency and the stress level of the simulations with the TMM model and the experiments match well. However, the TMM model using the Taylor assumption for the homogenization leads to a contradictory stress distribution. Both the stress level and course disagree the experimental results. Since the TMM model with the HS homogenization with a stiffness of the comparison material of $E_0 = 200\text{GPa}$ is close to the Taylor approximation, the discrepancy is likely due to the modeling of the diffusionless phase transformation via the "S"-shape model.

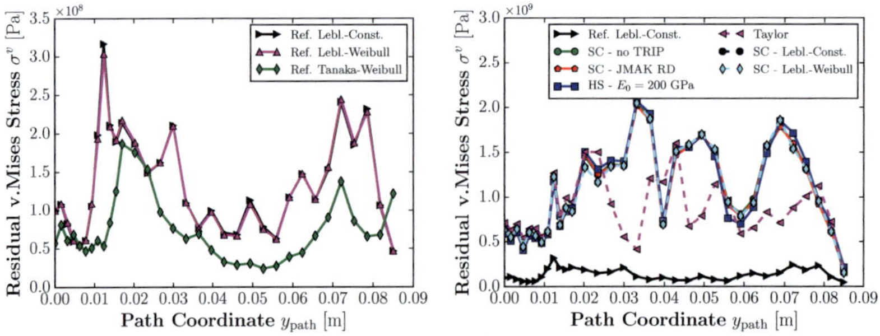

Figure 8.13: Residual v. Mises stress at the Path S of the reference model (left) and the TMM model (right) (Neumann et al., 2017)

A greater difference between the reference models and the TMM models becomes evident by considering the residual stress distribution along Path S (cf. Figure 8.9). The v. Mises stress curves of these different approaches differ strongly from each other with a factor of up to 10. Since the reference models predict a maximum v. Mises stress of about

300MPa, the maximum stress in the simulations with the TMM models is about 2000MPa (cf. Figure 8.13).

Summary and Conclusions

Considering the dilatation tests of the steel grade 22MnB5, in contrast to a direct implementation, the incremental implementation of the homogenization scheme and the phases' material behavior leads to a significant influence of the homogenization scheme on the overall material behavior in the thermo-elastic case with transformation strains. The influence of the modeling approaches in view of the TRIP effect, phase transformation behavior, and the homogenization scheme are considered in the hot stamping process of a w-shaped 22MnB5 sheet metal. Therefore the proposed TMM model is compared to both a phenomenological macroscopic model usually used in literature to simulate the hot stamping process and experimental results. It is found that the

- Resulting shapes of the simulations with both the TMM models and the reference models with each specification in view of the TRIP model, used phase transformation model, and homogenization scheme are in good agreement with the experimental results.
- Differences between the TMM models and the reference models become evident by considering the residual stresses.
- Stress dependence of the TRIP parameter has a minor influence on the residual stress prediction, whereas the type of the TRIP saturation function a significant one.
- Tanaka's TRIP saturation function leads to residual stresses which strongly differ from experimental results. However, Leblond's saturation function captures the tendency of the residual stress course but overestimates the stress level significantly using the reference models.

- Stress dependence of the TRIP parameter fitted with the Weibull function leads to a better agreement with experiments than the stress-independent TRIP parameter approach.
- Resulting residual stresses gained with the TMM models are in good agreement with the experimental ones in view of tendency of the stress course and stress level. To capture the nonlinear residual stress distribution, simulations with a finer discretization of the sheet metal thickness have to be performed.
- "S"-shape model for the description of the diffusionless martensitic phase transformation leads to a residual stress distribution with an opposite course throughout the thickness of the hot stamped sheet metal.

Chapter 9

Summary and Outlook

This work discusses the two-scale modeling of thermomechanical processing of steel, in particular, that of sheet metal in the context hot stamping. The characteristic of such processes is the phase transformation of the polycrystalline parent microstructure into several phases under thermomechanical loads. Thereby, thermal effects as the consumption or release of heat during the phase transformation and mechanical effects as the TRIP effect have to be considered. Within this work, great focus is put on the thermodynamically consistent modeling of the thermo-elasto-plastic phase behavior, the extension of the nonlinear homogenization scheme of HS-type, and the efficient numerical implementation of the resulting constitutive laws and localization relations into the commercial FE code ABAQUS/Standard. Furthermore, evolution laws describing the phase transformation effect are introduced into the homogenization scheme leading along with the TRIP effect to a strong coupling of the mechanical and thermal IBVP.
As a starting point for the modeling, experimental results of the hot stamping process of a w-shape and a u-shape demonstrator geometry have been used. Hot deep drawing processes of austenitized sheet metals of the steel grade 22MnB5 were performed at the IUL in Dortmund and the influence of the process parameters have been investigated. It was found that the influence of the process parameters on the shape is small. The process parameters do not appear to have any systematic influence on the residual stresses. Furthermore, the microstructure

contains ferritic, pearlitic, bainitic, and martensitic phases. The two latter observations can be explained by the fact that no cooling system was available to ensure standardized initial conditions for each forming process, as well as to produce a pure martensitic microstructure in the final part after quenching.

In tensile tests in accordance with the DIN EN 10002, both the initial and the final mechanical properties of the sheet metals were investigated. It was found that the initial and thermomechanical treated sheet metals are mainly mechanically isotropic and that the quenching rate and the holding time in the closed press have the greatest influence on the final mechanical state.

Based on three experimentally observable assumptions, a thermodynamically consistent bulk model for the phases has been found by introducing a Helmholtz free energy. The linear thermal expansion, the v. Mises plasticity, a modified Leblond's ansatz for the TRIP and an isotropic transformation strain tensor were introduced as eigenstrains. In literature and in this work, it is assumed that the GJ effect has a major and the Magee effect a minor influence in hot stamping processes. The GJ effect is modelled by a modified Leblond's ansatz. Kubler et al. (2011) proposed a transformation strain model taking both non-classical plasticity effects into account. The suggested TMM model can be easily extended by Kubler's ansatz.

Viscous effects, as viscoelastictiy and viscoplasticity, have been neglected in this work and should be considered in following approaches due to the high forming rate of the austenitic phase at high temperatures. With that, the overstresses in the forming step and relaxation effects in the quenching step can be taken into account in a more accurate way (see, e.g., Uppaluri and Helm, 2016).

In the models from the literature, the phase transformation has no impact on the evolution of the internal variable. For this reason, the assumption is tacitly done, that the dislocation density is untouched by the phase transformation. This strong limitation of these kind of models does not

hold for the suggested two-scale model. In contrast to the models in the literature, the TMM model introduces for each phase an internal variable which describes the plasticity in the associated phase and which can be modified by transformation effects.
Using the eigenstrain concept, the mechanical scale transition from the macro to the micro scale is done by a nonlinear extension of the HS homogenization method. The shortcoming of a stress response that is too stiff due to the description of the plastic strain of a phase by the mean value (c.f. the Transformation Field Analysis), is overcome by the possibility of adjusting the homogeneous comparison stiffness between the first-order bounds by Sachs and Taylor. However, the homogeneous stiffness tensor of the comparison material is set to fit the experimental findings best, e.g. the dilatation tests.
An essential result in this work in view of computational cost reduction is that, for an isotropic material behavior of the phases, the HS localization scheme is completely described by scalar parameters. The number of the parameters is twice the number of the phases in the microstructure. For the thermal scale transition, the well-known HS scheme is used, which can also be described by scalar localization factors for isotropic phases' thermal conductivities. The number of the localization factors is equal to the number of phases. Due to the similar phases' conductivities, the type of the thermal homogenization is of minor importance.
In experimental investigations, it was observed, that the specific heat capacities of the phases are temperature-dependent. This effect is not captured in the present TMM model and can be considered in further studies with respect to the formulation of the Helmholtz free energy and the thermodynamic consistency of the resulting model.
For the description of the diffusionless phase transformation, two models have been considered, the well-known KM model and an "S"-shape model. While the steel grade 42CrMo4 shows the "S"-shape behavior in dilatation tests, the steel grade 22MnB5 is exponential-like and captured well by the KM model. However, the stress impact on both the trans-

formation kinetics and the martensite formation start temperature are neglected in the present models.
The same is true for the diffusion-driven phase transformation. The JMAK and KV model were considered and extended to capture the cooling rate influence on the transformation kinetics. The stress influence is usually neglected. Both models can be easily modified by introducing a stress-dependent factor in the JMAK parameters or influence function in the KV model. However, for the parameter identification of such extended models, TTT diagrams and/or CCT diagrams determined under external mechanical load are needed (see, e.g., Min et al., 2012).
The effect of strain-induced transformation (STRIP) is also neglected in the considerations of the stress-transformation interaction. Min et al. (2013) found that the isothermal deformation at different temperatures above the MS temperature leads to deformation-induced ferrite transformation (DIFT) and deformation-induced bainite transformation (DIBT). The DIFT and DIBT effect significantly influences the thermomechanical properties of the final parts.
Furthermore, observations in experiments showed that the diffusion-driven phase transformation is dependent on both the austenitization temperature and time. The austenitization temperature and time have a direct influence on the austenitic grain size which influences the impact of the interface and volume diffusion-controlled growth of nucleated germs. This effect can be taken into account by Bok's grain size extension of the KV model.
The phase transformation models are usually valid for isothermal processes. In the non-isothermal process, Scheil's additivity rule provides a possibility of finding an evolution law for the volume fraction using the JMAK model or using the step-wise constant temperature method for the JMAK model, which is numerically efficient. Both ways to describe the non-isothermal are compared and slight differences between the methods are observed.
The numerical implementation was realized in ABAQUS/Standard us-

ing the subroutines UMAT, UMATHT, and UEXPAN. Due to the neglect of the effect of the mechanical dissipation on the heat equation and the neglect of the influence of the stress state on the phase transformation, a separate implementation of the heat equation and linear momentum equation can be performed. The finite deformation taking place in the deep drawing process is realized by the ULF leading to the incremental form of the phases' constitutive law and localization relation.

The parameter identification is performed for both the steel grades 42CrMo4 and 22MnB5. The elastic and the thermal parameters of the different phases were taken from literature. For the temperature dependent isotropic plastic hardening of 42CrMo4 and 22MnB5, a Voce-type and a Swift-type law have been used, respectively. The thermal expansion parameters are obtained from dilatation tests. From dilatation tests, one also obtains the transformation strains which have been modeled to be isotropic. The TRIP strains of the different phases are determined from dilatation tests under external loads applied after the heating of the specimen. In a completed phase transformation, the parameters of Leblond's model are identified.

The JMAK and the KV parameters are identified using the TTT diagram of the associated steel. For the steel 42CrMo4, the implementation of the JMAK model in the non-isothermal case, using the isothermal step method, was validated. The phase transformation model is validated by computing the CCT diagrams. While for the steel 42CrMo4, the transformation kinetics seem to fit well and the critical cooling rate deviates significantly from the experimental result. For the steel 22MnB5, the effect is the opposite. The model fits the critical cooling rates quite well, but the incubation time is too long in the model. Extensions of Scheil's rule suggested by, e.g., Réti and Felde (1999) can be used to get more accurate approximations.

The thermomechanical coupling is validated via dilatation tests. It was found that, in an incremental implementation of the TMM model, the impact of the homogenization scheme has a higher significance when compared to a direct implementation of the governing equations.

Finally, the hot stamping process of a w- and u-shaped part is simulated and the geometry and the residual stresses are compared to experimental and/or numerical results of the reference model. Note, that the transfer of the heated sheet metals from the furnace to the press is neglected. In the model, the die, blank holder, punch, and counter punch are isothermal rigid bodies, which neglect both the heating of the tools and the heat conductance through the tools. The heat transfer throughout the contact areas is set to be constant.

Considering the hot stamping of the w- and u-shaped geometries, it was found that the resulting shapes of the simulations with both the TMM models and the reference models are in good agreement with the experimental results. The differences between the TMM models and the reference models become evident when considering the residual stresses. It was observed that the stress dependence of the TRIP parameter has a minor influence, while the type of the TRIP saturation function has a significant influence on the residual stress distribution. In the reference models, Leblond's saturation function captures the tendency of the residual stress course but overestimates the stress level significantly. Furthermore, the stress dependence of the TRIP parameter fitted with the Weibull function leads to a better agreement with experiments than the stress-independent TRIP parameter approach. On the other hand, the resulting residual stresses gained with the TMM models are in good agreement with the experimental findings in view of tendency of the stress course and stress level. To capture the nonlinear residual stress distribution, simulations with a finer discretization of the sheet metal thickness have to be performed.

Appendix A

Specifications of the Deep Drawing Press

The specifications of the deep drawing press which has been used for the hot stamping experiments (see Figure 2.2) are given in Table A.1.

Table A.1: Technical data of the deep drawing press

Punch	Compression force		1000kN
	Retraction force		125kN
	Lift		600mm
	Lift velocity	downwards	350mm/s
		under load	20 – 43mm/s
		upwards	330mm/s
Drawing cushion	Compression force		250kN
	Lift		200mm
Engine data	Machine weight approx.		12250kg
	Tool weight max.		1500kg
	Weight of the hydraulic oil approx.		510kg
Power	Installed power approx.		39.5kW
	Current consumption approx.		73A
	Voltage		400V
	Frequency		50Hz

Appendix B

Summary of Results of Homogenization Theory

B.1 Eshelby-Solution based Homogenization Estimates

Many homogenization methods are based on the solution of the single inclusion problem. To be able to consider the single inclusion problem, the solution of the eigenstrain problem in a homogeneous material is needed.

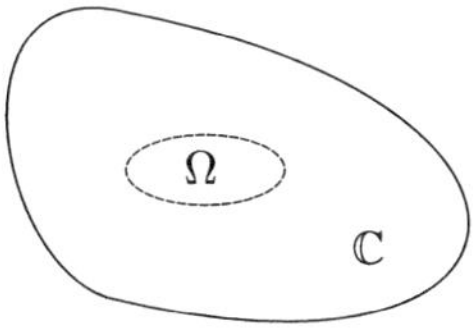

Figure B.1: Homogeneous eigenstrain problem

Assuming a homogeneous material with an ellipsoidal inclusion Ω with eigenstrain as depicted in Figure B.1, Eshelby (1957) found that the strain in Ω is constant. Furthermore, the strain can be determined by the linear relation

$$\varepsilon = \mathbb{E}\left[\varepsilon^{E}\right], \tag{B.1}$$

where $\boldsymbol{\varepsilon}^E$ denotes the eigenstrain tensor. The stress free eigenstrains $\boldsymbol{\varepsilon}^E$ can result from thermal or plastic strain in the inclusion. The fourth-order tensor $\mathbb{E}$ is referred to as Eshelby tensor and reads

$$\mathbb{E} = \frac{1+\nu}{3(1-\nu)}\mathbb{P}_1 + \frac{2(4-5\nu)}{15(1-\nu)}\mathbb{P}_2 \tag{B.2}$$

for spherical inclusions and isotropic elastic properties. Using Hooke's law, the uniform stress in Ω reads

$$\boldsymbol{\sigma} = \mathbb{C}\left[\boldsymbol{\varepsilon} - \boldsymbol{\varepsilon}^E\right] = \mathbb{C}\left(\mathbb{E} - \mathbb{I}^s\right)\left[\boldsymbol{\varepsilon}^E\right]. \tag{B.3}$$

Considering an elastically inhomogeneous inclusion problem (see left-hand side of Figure B.2), the problem has to be transferred to an equivalent homogeneous problem (Gross and Seelig, 2002) (see right-hand side of Figure B.2) for the determination of the strain in the inclusion due to the overall strain $\bar{\boldsymbol{\varepsilon}}$. The stress in the inclusion and the stress in the

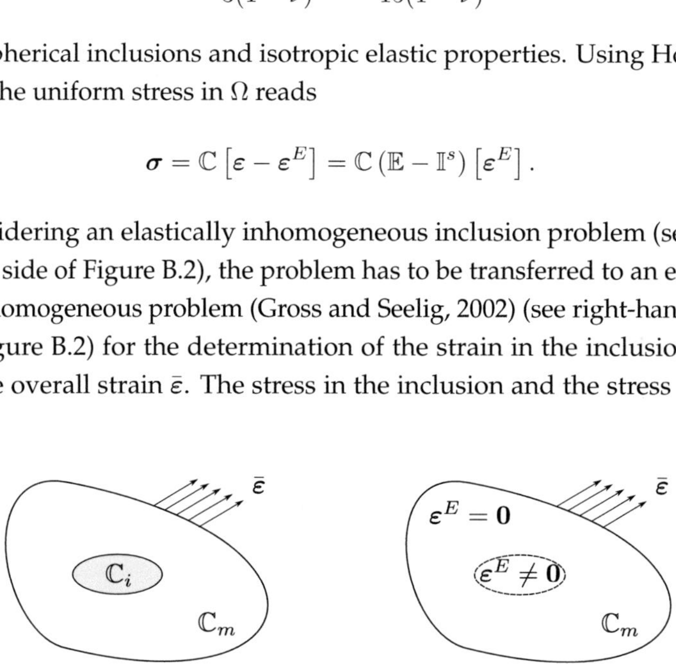

Figure B.2: Single inclusion problem and equivalently homogenized eigenstrain problem

eigenstrain area of the homogeneous problem read

$$\boldsymbol{\sigma}_i = \mathbb{C}_i\left[\boldsymbol{\varepsilon}_i\right] = \mathbb{C}_i\left[\bar{\boldsymbol{\varepsilon}} + \boldsymbol{a}\right] \tag{B.4}$$

and

$$\boldsymbol{\sigma}_i = \mathbb{C}_m\left[\boldsymbol{\varepsilon}_i - \boldsymbol{\varepsilon}^E\right] = \mathbb{C}_m\left[\bar{\boldsymbol{\varepsilon}} + \boldsymbol{a} - \boldsymbol{\varepsilon}^E\right], \tag{B.5}$$

respectively. Thereby, $\mathbb{C}_i$ denotes the stiffness in the inclusion, $\mathbb{C}_m$ the stiffness of the matrix, $\boldsymbol{a}$ the mismatch strain in the inclusion and $\boldsymbol{\varepsilon}_i$ the total strain in the inclusion. The mismatch strain is related to the eigenstrain via Eshelby's relation (B.1)

$$\boldsymbol{a} = \mathbb{E}\left[\boldsymbol{\varepsilon}^E\right] \tag{B.6}$$

leading to

$$\boldsymbol{\varepsilon}^E = -\left(\mathbb{E} + \left(\mathbb{C}_i - \mathbb{C}_m\right)^{-1}\mathbb{C}_m\right)^{-1}\left[\bar{\boldsymbol{\varepsilon}}\right] \tag{B.7}$$

via the equivalence of the stresses in both problems. The strain in the inclusion of the single inclusion problem is obtained by

$$\boldsymbol{\varepsilon}_i = \boldsymbol{\varepsilon} + \bar{\boldsymbol{\varepsilon}} = \mathbb{E}\left[\boldsymbol{\varepsilon}^E\right] + \bar{\boldsymbol{\varepsilon}} = \left(\mathbb{I}^s + \mathbb{E}\mathbb{C}_m^{-1}\left(\mathbb{C}_i - \mathbb{C}_m\right)\right)^{-1}\left[\bar{\boldsymbol{\varepsilon}}\right] = \mathbb{A}^{sip}\left[\bar{\boldsymbol{\varepsilon}}\right], \tag{B.8}$$

where $\mathbb{A}^{sip}$ is referred to as strain influence, strain concentration, or strain localization tensor of the single inclusion problem. Using the stress strain relation (B.4), the stress localization relation reads

$$\boldsymbol{\sigma}_i = \mathbb{C}_i\mathbb{A}^{sip}\mathbb{C}_m^{-1}\left[\bar{\boldsymbol{\sigma}}\right] = \mathbb{B}^{sip}\left[\bar{\boldsymbol{\sigma}}\right], \tag{B.9}$$

with the stress load $\mathbb{C}_m[\bar{\boldsymbol{\varepsilon}}]$. $\mathbb{B}^{sip}$ is referred to as stress influence, stress concentration, or stress localization tensor.

Using the localization relations of the single inclusion problem, a homogenization scheme for a dilute inclusion distribution can be found. Assuming a RVE with an interaction free distribution of the inclusions, the localization relation of the macroscopic stress or strain can be described by (B.8) or (B.9), respectively.

In literature, there are many other homogenization schemes describing the relation between the macroscopic quantities and the local ones. Some are discussed in the following.

Mori-Tanaka Approximation

The Mori-Tanaka approach assumes that the localization relation of the single inclusion problem (B.8) is valid between the matrix and the inclusion

$$\varepsilon_i = \mathbb{A}^{sip} \left[\varepsilon_m\right]. \tag{B.10}$$

Assuming a RVE with an interaction free distribution of the inclusions, the localization relation of the macroscopic strain

$$\varepsilon_i = \left(c_i \mathbb{I}^s + c_m \, {\mathbb{A}^{sip}}^{-1}\right)^{-1} \left[\bar{\varepsilon}\right] = \mathbb{A}^{MT} \left[\bar{\varepsilon}\right] \tag{B.11}$$

can be found, where the relation $\bar{\varepsilon} = c_i \varepsilon_i + c_m \varepsilon_m$ has been used. For a multi inclusion material, i.e., a matrix with $N-1$ different phases, Benveniste (1987) interpreted the extension of Mori-Tanaka's theory

$$\mathbb{A}_\alpha^{MT} = \left(c_\alpha \mathbb{I}^S + c_m \, {\mathbb{A}_\alpha^0}^{-1} + \sum_{\beta=2}^{N} c_\beta \mathbb{A}_\beta^0 \, {\mathbb{A}_\alpha^0}^{-1}\right)^{-1} \tag{B.12}$$

in that way that each phase behaves like an isolated single inclusion in the matrix with the far-field strain ε_m. The tensor $\mathbb{A}_\alpha^0$ reads

$$\mathbb{A}_\alpha^0 = \left(\mathbb{I}^s + \mathbb{E}\mathbb{C}_m^{-1} \left(\mathbb{C}_\alpha - \mathbb{C}_m\right)\right)^{-1}. \tag{B.13}$$

The localization tensor for the matrix is deducible via condition $\langle \mathbb{A}^{MT} \rangle = \mathbb{I}^s$

$$\mathbb{A}_m^{MT} = \frac{1}{c_m} \left(\mathbb{I}^s - \sum_{\alpha=2}^{N} c_\alpha \mathbb{A}_\alpha^{MT}\right), \tag{B.14}$$

where the matrix is set to be phase one.

Self-Consistent Scheme with a Distinct Phase

In the self-consistent approximation, it is assumed that the inclusion is embedded in an infinite volume of an effective medium with stiffness

$\bar{\mathbb{C}} = \langle \mathbb{C}\mathbb{A} \rangle$. Due to this, the strain localization tensor reads

$$\mathbb{A}^{SC} = \left(\mathbb{I}^s + \mathbb{E}\bar{\mathbb{C}}^{-1}\left(\mathbb{C}_i - \bar{\mathbb{C}}\right)\right)^{-1} \tag{B.15}$$

$$= \left(\mathbb{I}^s + \mathbb{E}\left\langle \mathbb{C}\mathbb{A}^{SC}\right\rangle^{-1}\left(\mathbb{C}_i - \left\langle \mathbb{C}\mathbb{A}^{SC}\right\rangle\right)\right)^{-1}. \tag{B.16}$$

This is a nonlinear equation for the determination of the strain localization tensor $\mathbb{A}^{SC}$.

For a multi phase material with a distinct phase, which is called matrix, the strain localization tensors for the inclusions read

$$\mathbb{A}_\alpha^{SC} = \left(\mathbb{I}^s + \mathbb{E}\left\langle \mathbb{C}\mathbb{A}^{SC}\right\rangle^{-1}\left(\mathbb{C}_\alpha - \left\langle \mathbb{C}\mathbb{A}^{SC}\right\rangle\right)\right)^{-1}. \tag{B.17}$$

This is a set of nonlinear equations for the strain localization tensor of the inclusions $\mathbb{A}_\alpha$ (Willis, 1977). The strain localization tensor of the distinct phase, i.e., the matrix,

$$\mathbb{A}_m^{SC} = \frac{1}{c_m}\left(\mathbb{I}^s - \sum_{\alpha=2}^{N} c_\alpha \mathbb{A}_\alpha^{SC}\right), \tag{B.18}$$

is the result of the condition $\left\langle \mathbb{A}^{SC}\right\rangle = \mathbb{I}^s$.

The Variational Principle of Hashin-Shtrikman

Introducing a homogeneous linear comparison material with a stiffness denoted by $\mathbb{C}_0$, the difference of the stresses between the initial problem and the problem with the homogeneous comparison material due to real strain ε

$$\boldsymbol{p} = \boldsymbol{\sigma} - \mathbb{C}_0\left[\varepsilon\right] = \left(\mathbb{C} - \mathbb{C}_0\right)\left[\varepsilon\right] = \delta\mathbb{C}\left[\varepsilon\right] \tag{B.19}$$

is defined and referred to as stress polarization. Thus, the static BVP

$$\operatorname{div}\left(\boldsymbol{\sigma}\right) = \mathbf{0}, \qquad \langle \varepsilon \rangle = \bar{\varepsilon} \tag{B.20}$$

can be re-written into the equivalent BVP

$$\operatorname{div}\left(\mathbb{C}_0\left[\varepsilon\right]\right) + \boldsymbol{f} = \mathbf{0}, \qquad \langle\varepsilon\rangle = \bar{\varepsilon}. \tag{B.21}$$

By this procedure, the initial inhomogeneous BVP without body force (B.20) is transformed into a homogeneous BVP with the body force $\boldsymbol{f} = \operatorname{div}(\boldsymbol{p})$ (c.f. equation (B.21)). Assuming the stress polarization to be known, the solution of the BVP (B.20) formally reads

$$\varepsilon = \bar{\varepsilon} - \mathcal{G}\left[\boldsymbol{p} - \langle\boldsymbol{p}\rangle\right], \tag{B.22}$$

where the integral operator

$$\mathcal{G}\left[\boldsymbol{p} - \langle\boldsymbol{p}\rangle\right] = \int_{V'} \boldsymbol{\Gamma}\left(\boldsymbol{x} - \boldsymbol{x}'\right)\left[\boldsymbol{p} - \langle\boldsymbol{p}\rangle\right]\, \mathrm{d}V' \tag{B.23}$$

includes the second derivatives of infinite body's Green's function in $\boldsymbol{\Gamma}$ (Zeller and Dederichs, 1973; Willis, 1977). For the special case of a polycrystal with phase wise constant material properties and phase wise constant stress polarization, the integral operator is a fourth-order tensor which depends on the Eshelby tensor and the homogeneous stiffness

$$\mathbb{P}_0 = \mathbb{E}\mathbb{C}_0^{-1}. \tag{B.24}$$

The formal solution (B.22) is given by the N equations

$$\varepsilon_\alpha = \bar{\varepsilon} - \mathbb{P}_0\left[\boldsymbol{p}_\alpha - \langle\boldsymbol{p}\rangle\right], \tag{B.25}$$

where the stress polarization of phase α is given by $\boldsymbol{p}_\alpha = \boldsymbol{\sigma}_\alpha - \mathbb{C}_0\left[\varepsilon_\alpha\right]$. With Hooke's law $\boldsymbol{\sigma}_\alpha = \mathbb{C}_\alpha\left[\varepsilon_\alpha\right]$, the solution can be re-written into the phase interaction law for the determination of the phase stresses or strains

$$\boldsymbol{\sigma}_\alpha - \bar{\boldsymbol{\sigma}} = \mathbb{L}\left[\varepsilon_\alpha - \bar{\varepsilon}\right], \tag{B.26}$$

where Hill's constraint tensor is denoted by $\mathbb{L} = \mathbb{C}_0 - \mathbb{P}_0^{-1}$. The strain localization relation

$$\varepsilon_\alpha = \left(\mathbb{P}_0^{-1} + \delta\mathbb{C}_\alpha\right)^{-1} \left\langle \left(\mathbb{P}_0^{-1} + \delta\mathbb{C}\right)^{-1} \right\rangle^{-1} [\bar{\varepsilon}] = \mathbb{A}_\alpha^{HS} [\bar{\varepsilon}] \tag{B.27}$$

fulfills the phase interaction law (B.26), where $\delta\mathbb{C}_\alpha = \mathbb{C}_\alpha - \mathbb{C}_0$ holds. The strain localization tensor $\mathbb{A}_\alpha^{HS}$ is a function of the homogeneous stiffness $\mathbb{C}_0$ which can be arbitrarily chosen. Considering equation (B.26), the first-order bounds of Reuss and Voigt are obtained for $\|\mathbb{L}\| \to 0$, if $\|\mathbb{C}_0\| \to 0$, and $\|\mathbb{L}\| \to \infty$, if $\|\mathbb{C}_0\| \to \infty$, respectively (Ponte Castañeda and Suquet, 1998; Böhlke et al., 2014). Tighter bounds, i.e., bounds of second-order, are obtained if the difference of the stiffness tensors $\delta\mathbb{C}_\alpha$ is negative or positive definite for all phases (Gross and Seelig, 2002). This can be shown by considering the difference of the effective strain energy of the homogeneous comparison material and the effective strain energy $\mathcal{F}^m = W_0 - \bar{W} = \bar{\varepsilon} \cdot \mathbb{C}_0 [\bar{\varepsilon}] /2 - \bar{\varepsilon} \cdot \bar{\mathbb{C}} [\bar{\varepsilon}] /2$ (see, e.g., Willis, 1977). The difference determines the functional

$$\mathcal{F}^m = \left\langle \frac{1}{2}\boldsymbol{p} \cdot \delta\mathbb{C}^{-1} [\boldsymbol{p}] + \frac{1}{2}\boldsymbol{p} \cdot \mathbb{P}_0 [\boldsymbol{p}] - \boldsymbol{p} \cdot \bar{\varepsilon} \right\rangle, \tag{B.28}$$

for a polycrystal microstructure with piece-wise constant phase properties and stress polarization (Hashin and Shtrikman, 1962b). The stationary point of this functional is determined from $\delta\mathcal{F}^m = 0$ and is given by relation (B.25).

Self-Consistent Scheme without Distinct Phase

With the strain localization tensor from the variational principle of Hashin-Shtrikman

$$\mathbb{A}_\alpha^{HS}(\mathbb{C}_0) = \left(\mathbb{P}_0^{-1} + \delta\mathbb{C}_\alpha\right)^{-1} \left\langle \left(\mathbb{P}_0^{-1} + \delta\mathbb{C}\right)^{-1} \right\rangle^{-1}, \tag{B.29}$$

a self-consistent method without a distinct phase can be found. For this purpose, the homogeneous stiffness of the comparison material is chosen to be the effective stiffness of the microstructure

$$\mathbb{C}_0 = \bar{\mathbb{C}} = \left\langle \mathbb{C}\mathbb{A}^{HS}(\mathbb{C}_0) \right\rangle . \tag{B.30}$$

This condition is a set of nonlinear equations for the determination of the homogeneous stiffness of the comparison material $\mathbb{C}_0$.

B.2 First-Order Bounds of Thermal and Mechanical Properties

The thermal or mechanical first-order homogenization schemes can be derived by the consideration of the associated minimum and maximum energies. In an energetic point of view, the first-order schemes are bounds for the effective thermal or mechanical material behavior.

Thermal Energetic First-Order Bounds

The thermal energy of a phase α and the macroscopic thermal energy are given by

$$W_\alpha^g = \frac{1}{2} \boldsymbol{g}_\alpha \cdot \boldsymbol{\lambda}_\alpha \boldsymbol{g}_\alpha \tag{B.31}$$

and

$$\bar{W}^g = \langle W^g \rangle = \frac{1}{2} \langle \boldsymbol{g} \cdot \boldsymbol{\lambda} \boldsymbol{g} \rangle , \tag{B.32}$$

respectively. Assuming that the temperature gradients of the phases assemble the way that the macroscopic energy takes a minimum, the optimization problem

$$\mathcal{G} = [\tilde{\boldsymbol{g}}_1, \dots, \tilde{\boldsymbol{g}}_N]^{\mathrm{T}} = \underset{\boldsymbol{g}_\alpha, \forall \alpha \in \mathcal{PH}}{\arg} \left(\min \left(I_g \right) \right) \tag{B.33}$$

has to be solved, where

$$I_g = \left\{ \bar{W}^g(\mathcal{G}) \mid \mathcal{G} = [\boldsymbol{g}_1, \dots, \boldsymbol{g}_N]^{\mathrm{T}} ,\ \boldsymbol{g}_\alpha \in \mathcal{A}^\theta \ \forall \alpha \in \mathcal{PH};\ \langle \boldsymbol{g} \rangle = \bar{\boldsymbol{g}} \right\} \tag{B.34}$$

has been defined. The set of properly temperature gradients is denoted by $\mathcal{A}^\theta$. The principle of the energy minimum leads to the thermal Lagrange equation

$$\mathcal{L}_g = \bar{W}^g - \boldsymbol{\gamma} \cdot (\langle \boldsymbol{g} \rangle - \bar{\boldsymbol{g}}), \tag{B.35}$$

whose stationary point can be found by zeroing the gradient

$$\frac{\partial \mathcal{L}_g}{\partial \mathcal{G}} = \begin{bmatrix} \frac{\partial \mathcal{L}_g}{\partial \boldsymbol{g}_1} \\ \vdots \\ \frac{\partial \mathcal{L}_g}{\partial \boldsymbol{g}_N} \end{bmatrix} = \begin{bmatrix} c_1 \boldsymbol{\lambda}_1 \boldsymbol{g}_1 - c_1 \boldsymbol{\gamma} \\ \vdots \\ c_N \boldsymbol{\lambda}_N \boldsymbol{g}_N - c_N \boldsymbol{\gamma} \end{bmatrix} = \mathbb{O}, \tag{B.36}$$

where γ denotes the Lagrange parameter. This leads to the set of N linear equations

$$c_\alpha \boldsymbol{\lambda}_\alpha \boldsymbol{g}_\alpha - c_\alpha \boldsymbol{\gamma} = \boldsymbol{0} \quad \forall \alpha \in \mathcal{PH}, \tag{B.37}$$

where $N = |\mathcal{PH}|$ is the number of phases. The solution of this set of equations reads

$$\boldsymbol{\lambda}_\alpha \boldsymbol{g}_\alpha = \boldsymbol{q}_\alpha = \boldsymbol{\gamma} \tag{B.38}$$

leading to the Reuss-type heat flux localization relation

$$\boldsymbol{q}_\alpha = \bar{\boldsymbol{q}} \tag{B.39}$$

and via

$$\bar{\boldsymbol{q}} = \left\langle \boldsymbol{\lambda}^{-1} \right\rangle^{-1} \bar{\boldsymbol{g}} \tag{B.40}$$

to the macroscopic heat conductivity tensor

$$\bar{\boldsymbol{\lambda}}_l = \left\langle \boldsymbol{\lambda}^{-1} \right\rangle^{-1} = \left\langle \lambda^{-1} \right\rangle^{-1} \boldsymbol{I}. \tag{B.41}$$

This is the first first-order bound. In a similar procedure, the temperature gradient localization relation can be determined via the complementary thermal phase energy

$$W_\alpha^q = \frac{1}{2}\boldsymbol{q}_\alpha \cdot \boldsymbol{\lambda}_\alpha^{-1}\boldsymbol{q}_\alpha \tag{B.42}$$

and the complementary macroscopic thermal energy

$$\bar{W}^q = \langle W^q \rangle = \frac{1}{2}\left\langle \boldsymbol{q} \cdot \boldsymbol{\lambda}^{-1}\boldsymbol{q} \right\rangle. \tag{B.43}$$

The principle of minimum energy

$$[\tilde{\boldsymbol{q}}_1, \ldots, \tilde{\boldsymbol{q}}_N]^{\mathrm{T}} = \underset{\boldsymbol{q}_\alpha,\, \forall\alpha\in\mathcal{PH}}{\arg} \left(\min\left(I_2\right)\right) \tag{B.44}$$

with

$$I_q = \left\{ \bar{W}^q(\mathcal{Q}) \mid \mathcal{Q} = [\boldsymbol{q}_1, \ldots, \boldsymbol{q}_N]^{\mathrm{T}},\ \langle \boldsymbol{q} \rangle = \bar{\boldsymbol{q}} \right\} \tag{B.45}$$

leads to the Voigt-type temperature gradient localization relation

$$\boldsymbol{g}_\alpha = \bar{\boldsymbol{g}} \tag{B.46}$$

and the macroscopic heat conductivity tensor

$$\bar{\boldsymbol{\lambda}}_u = \langle \boldsymbol{\lambda} \rangle = \langle \lambda \rangle \boldsymbol{I}. \tag{B.47}$$

From the principle of minimum thermal energy, it follows that the macroscopic energy with the effective heat conductivity tensor is smaller than the volume average of the phases' thermal energy for arbitrary temperature gradients

$$\langle \boldsymbol{g} \cdot \boldsymbol{\lambda}\boldsymbol{g} \rangle \geq \bar{\boldsymbol{g}} \cdot \bar{\boldsymbol{\lambda}}\bar{\boldsymbol{g}}. \tag{B.48}$$

With the first-order approximation (B.46), the Voigt-type localization scheme appears to be the upper first-order bound for the effective material behavior from an energetic point of view

$$\bar{\boldsymbol{g}} \cdot \left(\bar{\boldsymbol{\lambda}}_u - \bar{\boldsymbol{\lambda}}\right) \bar{\boldsymbol{g}} \geq 0. \tag{B.49}$$

Considering the complementary thermal energy, it can be shown that the Reuss-type localization scheme is a lower first-order bound for the effective material behavior from an energetic point of view.

Mechanical Energetic First-Order Bounds

The mechanical energy of a phase α and the macroscopic mechanical energy are given by

$$\rho_\alpha \psi_\alpha^\varepsilon = W_\alpha^\varepsilon = \frac{1}{2}\left(\boldsymbol{\varepsilon}_\alpha - \boldsymbol{\varepsilon}_\alpha^E\right) \cdot \mathbb{C}_\alpha \left[\boldsymbol{\varepsilon}_\alpha - \boldsymbol{\varepsilon}_\alpha^E\right] \tag{B.50}$$

and

$$\bar{W}^\varepsilon = \left\langle W^E \right\rangle = \frac{1}{2}\left\langle \left(\boldsymbol{\varepsilon} - \boldsymbol{\varepsilon}^E\right) \cdot \mathbb{C}\left[\boldsymbol{\varepsilon} - \boldsymbol{\varepsilon}^E\right] \right\rangle, \tag{B.51}$$

respectively. Because of the following minimization of the macroscopic energy with respect to the phase strains, only the first part of the phases' Helmholtz free energy given in (5.17) is taken into account. This part is denoted by $\rho_\alpha \psi_\alpha^\varepsilon = W_\alpha^\varepsilon$ and usually referred to as strain energy.

Assuming the principle of minimum energy, the macroscopic strain energy has to take a minimum which leads to the optimization problem

$$\mathcal{E} = \left[\tilde{\boldsymbol{\varepsilon}}_1, \ldots, \tilde{\boldsymbol{\varepsilon}}_N\right]^{\mathrm{T}} = \underset{\boldsymbol{\varepsilon}_\alpha,\, \forall \alpha \in \mathcal{PH}}{\arg} \left(\min\left(I_\varepsilon\right)\right) \tag{B.52}$$

with definition of the set of possible macroscopic strain energies

$$I_\varepsilon = \left\{ \bar{W}^\varepsilon(\mathcal{E}) \,\middle|\, \mathcal{E} = \left[\boldsymbol{\varepsilon}_1, \ldots, \boldsymbol{\varepsilon}_N\right]^{\mathrm{T}},\ \boldsymbol{\varepsilon}_\alpha \in \mathcal{A}^\varepsilon\, \forall \alpha \in \mathcal{PH};\ \langle \boldsymbol{\varepsilon} \rangle = \bar{\boldsymbol{\varepsilon}} \right\}. \tag{B.53}$$

The set of properly strains is denoted by $\mathcal{A}^{\varepsilon}$. The principle of the energy minimum leads to the mechanical Lagrange equation

$$\mathcal{L}_{\varepsilon} = \bar{W}^{\varepsilon} - \boldsymbol{\gamma} \cdot (\langle \boldsymbol{\varepsilon} \rangle - \bar{\boldsymbol{\varepsilon}}), \tag{B.54}$$

whose stationary point can be found by zeroing the gradient

$$\frac{\partial \mathcal{L}_{\varepsilon}}{\partial \mathcal{E}} = \begin{bmatrix} \dfrac{\partial \mathcal{L}_{\varepsilon}}{\partial \boldsymbol{\varepsilon}_1} \\ \vdots \\ \dfrac{\partial \mathcal{L}_{\varepsilon}}{\partial \boldsymbol{\varepsilon}_N} \end{bmatrix} = \begin{bmatrix} c_1 \mathbb{C}_1 \left[\boldsymbol{\varepsilon}_1 - \boldsymbol{\varepsilon}_1^E\right] - c_1 \boldsymbol{\gamma} \\ \vdots \\ c_N \mathbb{C}_N \left[\boldsymbol{\varepsilon}_N - \boldsymbol{\varepsilon}_N^E\right] - c_N \boldsymbol{\gamma} \end{bmatrix} = \mathbb{O}. \tag{B.55}$$

The solution of this N equations, where $N = |\mathcal{PH}|$ denotes the number of phases,

$$\mathbb{C}_{\alpha} \left[\boldsymbol{\varepsilon}_{\alpha} - \boldsymbol{\varepsilon}_{\alpha}^E\right] = \boldsymbol{\sigma}_{\alpha} = \boldsymbol{\gamma} \tag{B.56}$$

corresponds to the Sachs stress localization scheme

$$\boldsymbol{\sigma}_{\alpha} = \bar{\boldsymbol{\sigma}}. \tag{B.57}$$

Volume averaging of the strain and use of the Sachs stress localization relation, leads to the Sachs strain localization relation

$$\boldsymbol{\varepsilon}_{\alpha} = \mathbb{A}_{\alpha}^S [\bar{\boldsymbol{\varepsilon}}] + \boldsymbol{\varepsilon}_{\alpha}^E - \mathbb{A}_{\alpha}^S \left[\langle \boldsymbol{\varepsilon}^E \rangle\right], \tag{B.58}$$

with the Sachs strain localization tensor

$$\mathbb{A}_{\alpha}^S = \mathbb{C}_{\alpha}^{-1} \left\langle \mathbb{C}^{-1} \right\rangle^{-1}. \tag{B.59}$$

Note, that the Sachs strain localization tensor corresponds to the Reuss strain localization tensor in the elastic case. The macroscopic stiffness tensor for the Sachs localization scheme is determined by

$$\bar{\mathbb{C}}^S = \left\langle \mathbb{C} \mathbb{A}^S \right\rangle = \left\langle \mathbb{C}^{-1} \right\rangle^{-1}. \tag{B.60}$$

In the special case of isotropic phase behavior, two separate strain localization relations, one for the spherical part

$$\varepsilon_\alpha^\circ = \tau_\alpha^S \bar{\varepsilon}^\circ + \varepsilon_\alpha^{E\circ} - \tau_\alpha^S \left\langle \varepsilon^{E\circ} \right\rangle , \tag{B.61}$$

and one for the deviatoric part

$$\varepsilon_\alpha' = \varsigma_\alpha^S \bar{\varepsilon}' + \varepsilon_\alpha^{E'} - \varsigma_\alpha^S \left\langle \varepsilon^{E'} \right\rangle , \tag{B.62}$$

can be found. The $2N$ scalar localization parameters are the limits for a vanishing stiffness of the homogeneous comparison material of the HS localization parameters (5.70) and read

$$\tau_\alpha^S = \frac{K_\alpha^{-1}}{\langle K^{-1} \rangle}, \qquad \varsigma_\alpha^S = \frac{G_\alpha^{-1}}{\langle G^{-1} \rangle}. \tag{B.63}$$

Considering the complementary strain energy, which is got by a Legendre transformation of the strain energy,

$$W_\alpha^\sigma = \frac{1}{2} \boldsymbol{\sigma}_\alpha \cdot \mathbb{C}_\alpha^{-1} [\boldsymbol{\sigma}_\alpha] + \boldsymbol{\sigma}_\alpha \cdot \boldsymbol{\varepsilon}_\alpha^E , \tag{B.64}$$

the minimum of complementary energy is obtained by the optimization problem

$$\mathcal{S} = [\tilde{\boldsymbol{\sigma}}_1, \ldots, \tilde{\boldsymbol{\sigma}}_N]^{\mathrm{T}} = \underset{\boldsymbol{\sigma}_\alpha, \forall \alpha \in \mathcal{PH}}{\arg} \left(\min \left(I_\sigma \right) \right) \tag{B.65}$$

with definition of the set of possible macroscopic complementary strain energies

$$I_\sigma = \left\{ \bar{W}^\sigma(\mathcal{S}) \mid \mathcal{S} = [\boldsymbol{\sigma}_1, \ldots, \boldsymbol{\sigma}_N]^{\mathrm{T}} , \; \boldsymbol{\sigma}_\alpha \in \mathcal{A}^\sigma \, \forall \alpha \in \mathcal{PH}; \; \langle \boldsymbol{\sigma} \rangle = \bar{\boldsymbol{\sigma}} \right\} . \tag{B.66}$$

The set of properly phase stresses is denoted by $\mathcal{A}^\sigma$ and the macroscopic complementary strain energy is given by

$$\bar{W}^\sigma = \langle W^\sigma \rangle . \tag{B.67}$$

The introduction of a Legendre function and search for the stationary point leads to both the Taylor strain localization scheme

$$\varepsilon_\alpha = \bar{\varepsilon} \tag{B.68}$$

and the macroscopic stiffness tensor

$$\bar{\mathbb{C}}^T = \langle \mathbb{C} \rangle . \tag{B.69}$$

The corresponding Taylor stress localization tensor reads

$$\boldsymbol{\sigma}_\alpha = \mathbb{B}_\alpha^T [\bar{\boldsymbol{\sigma}}] + \mathbb{B}_\alpha^T \left[\langle \mathbb{C} \left[\varepsilon^E \right] \rangle \right] - \varepsilon_\alpha^E \tag{B.70}$$

with the Taylor stress localization tensor

$$\mathbb{B}_\alpha^T = \mathbb{C}_\alpha \langle \mathbb{C} \rangle^{-1} . \tag{B.71}$$

For the special case of isotropic material behavior, the stress localization relation can be reduced to two separate equations

$$\boldsymbol{\sigma}_\alpha^\circ = \frac{K_\alpha}{\langle K \rangle} \bar{\boldsymbol{\sigma}}^\circ + 3 \frac{K_\alpha}{\langle K \rangle} \left[\langle K \left[\varepsilon^{E\circ} \right] \rangle \right] - \varepsilon_\alpha^{E\circ} , \tag{B.72}$$

$$\boldsymbol{\sigma}'_\alpha = \frac{G_\alpha}{\langle G \rangle} \bar{\boldsymbol{\sigma}}' + 2 \frac{G_\alpha}{\langle G \rangle} \left[\left\langle G \left[\varepsilon^{E'} \right] \right\rangle \right] - \varepsilon_\alpha^{E'} . \tag{B.73}$$

With the same arguments which has been used above to show that the thermal first-order schemes are bounds for the effective thermal material behavior, it can be shown that the Sachs localization scheme is the lower and the Taylor localization scheme is the upper bound for the effective mechanical behavior from an energetic point of view.

Appendix C

Influence of Mechanical Loads on the Phases' Volume Fraction and Density

Isotropic mechanical and thermal loads lead to a volume change of a bulk. In a microstructure, which consists of different phases with

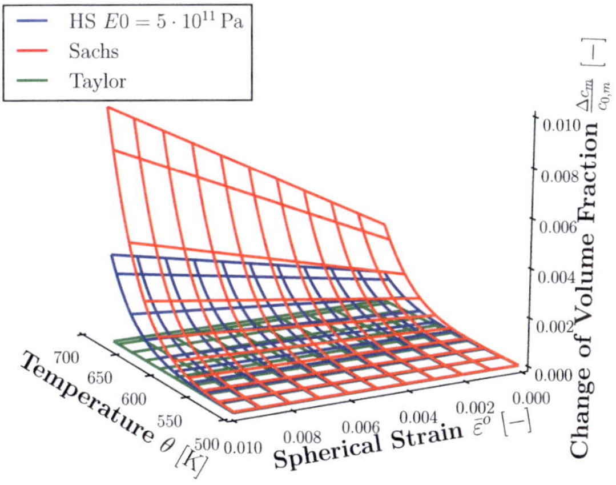

Figure C.1: Change of the volume fraction of martensite due to spherical effective strain

differing bulk moduli or thermal expansion coefficients, the volume fractions change because of different volume increases of each phase.

For a phase α, the volume fraction at the end of the loading process reads

$$c_{1,\alpha} = c_{0,\alpha} \left(1 + \sum_{\beta \in \mathcal{PH}, \beta \neq \alpha} c_{0,\beta} \mathrm{tr} \left(\boldsymbol{\varepsilon}_\alpha - \boldsymbol{\varepsilon}_\beta \right) \right), \tag{C.1}$$

where $c_{0,\alpha}$ denotes the initial volume fraction. Note, that assuming a Voigt or Taylor homogenization scheme, the volume fractions of all phases at the end of the loading process are equal to the initial volume fractions.

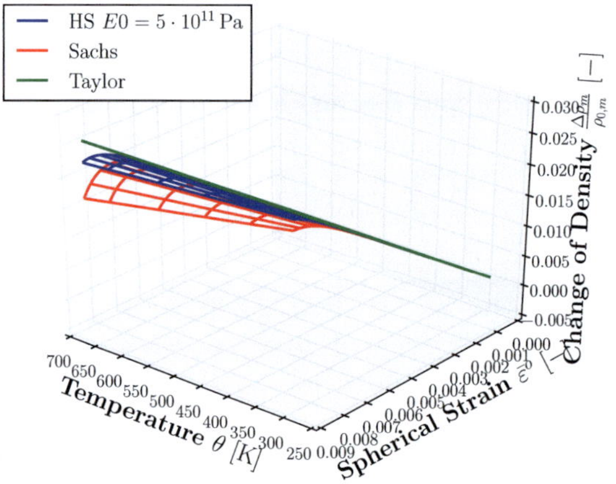

Figure C.2: Change of the density of martensite due to spherical effective strain

With the HS localization relation for the spherical part of the phase strain (5.71), the relation reads

$$\Delta c_\alpha = c_{0,\alpha} \sum_{\beta \in \mathcal{PH}, \beta \neq \alpha} c_{0,\beta} \left\langle \frac{\tau}{K} \right\rangle^{-1} \left(\mathrm{tr} \left(\Delta_\beta \left(\bar{\boldsymbol{\varepsilon}}^\circ - \left\langle \tau \boldsymbol{\varepsilon}^{E\circ} \right\rangle \right) + \left\langle \frac{\tau}{K} \right\rangle \left(\tau_\alpha \boldsymbol{\varepsilon}_\alpha^{E\circ} - \tau_\beta \boldsymbol{\varepsilon}_\beta^{E\circ} \right) \right) \right), \tag{C.2}$$

where the change of the phases' volume fraction due to spherical load $\Delta c_\alpha = c_{1,\alpha} - c_{0,\alpha}$ and the difference $\Delta_\beta = \tau_\alpha / K_\alpha - \tau_\beta / K_\beta$ have been introduced. For an austenitic-martensitic microstructure of the steel grade 22MnB5 with the parameters given in Table 7.5, the change of volume fraction due to spherical load is shown on the left-hand side of Figure C.1 dependent on the temperature and spherical effective strain. The volume fractions have been determined by the KM law (4.107). The HS solution is always inbetween the regime spanned by the first-order bounds by Taylor and Sachs. The maximum volume fraction change due to spherical loads is about 1% for the Sachs homogenization scheme at a spherical load of about $\varepsilon^\circ = 0.01$ and, therefore, can be neglected in thermomechanical studies.

The same analysis can be done for the density, resulting in

$$\rho_{1,\alpha} = \frac{|v_{1,\alpha}|}{|v_{0,\alpha}|} \rho_{0,\alpha} = \frac{|v_{0,\alpha}| + |\Delta v_\alpha|}{|v_{0,\alpha}|} \rho_{0,\alpha} = \rho_{0,\alpha} \left(1 + \operatorname{tr}\left(\boldsymbol{\varepsilon}_\alpha\right)\right) \tag{C.3}$$

and

$$\Delta \rho_\alpha = \rho_{1,\alpha} - \rho_{0,\alpha} = \rho_{0,\alpha} \left\langle \frac{\tau}{K} \right\rangle^{-1} \frac{\tau_\alpha}{K_\alpha} \operatorname{tr}\left(\bar{\boldsymbol{\varepsilon}}^\circ + K_\alpha \left\langle \frac{\tau}{K} \right\rangle \boldsymbol{\varepsilon}_\alpha^{E\circ} - \left\langle \tau \boldsymbol{\varepsilon}^{E\circ} \right\rangle \right) \tag{C.4}$$

with the HS strain localization relations. For an austenitic-martensitic microstructure of the steel grade 22MnB5, the change of density due to spherical load is depicted on the right-hand side of Figure C.2 with an initial density of $\rho_{0,\alpha} = 7850 \mathrm{kg/m}^3$. The change of the density is linear in the spherical effective strain with a maximum density change due to spherical loads of about 3% for the Sachs homogenization scheme at a spherical load of about $\varepsilon^\circ = 0.01$. The phases' density change is neglected in the thermomechanical studies.

In the following, relation (C.1), is derived. The volume of the microstructure at the end of the loading process is the sum of the phase volumes at the end of the process $|v_{1,ges}| = \sum_\alpha |v_{1,\alpha}|$. The volume fraction at the end of the loading reads

$$c_{1,\alpha} = \frac{|v_{1,\alpha}|}{|v_1^{ges}|} = \frac{|v_{0,\alpha}| + |\Delta v_\alpha|}{\sum\limits_{\beta \in \mathcal{PH}} (|v_{0,\beta}| + |\Delta v_\beta|)}, \tag{C.5}$$

where $|v_{0,\alpha]}|$ is the initial volume and $|\Delta v_\alpha|$ the increase of the volume of phase α after loading. This is a nonlinear equation for the determination of the volume increase of each phase on the microstructure

$$c_{1,\alpha} = c_{1,\alpha}\left(|\Delta v_1|, \ldots, |\Delta v_N|\right). \tag{C.6}$$

Since volume increments due to mechanical and thermal loads are small, a linearization

$$c_{1,\alpha} = c_{0,\alpha} + \sum_{\beta \in \mathcal{PH}} \left. \frac{\partial c_{1,\alpha}}{\partial |\Delta v_\beta|} \right|_{|\Delta v_1|=0,\ldots,|\Delta v_N|=0} |\Delta v_\beta| \tag{C.7}$$

with the derivations

$$\frac{\partial c_{1,\alpha}}{\partial |\Delta v_\beta|} = \begin{cases} -\dfrac{|v_{0,\alpha}|+|\Delta v_\alpha|}{\left(\sum\limits_{\gamma \in \mathcal{PH}} (|V_{0,\gamma}|+|\Delta v_\gamma|)\right)^2} & \text{if} \quad \alpha \neq \beta \\ \dfrac{\sum\limits_{\gamma \in \mathcal{PH}} (|v_{0,\gamma}|+|\Delta v_\gamma|) - (|v_{0,\alpha}|+|\Delta v_\alpha|)}{\left(\sum\limits_{\gamma \in \mathcal{PH}} (|v_{0,\gamma}|+|\Delta v_\gamma|)\right)^2} & \text{if} \quad \alpha = \beta \end{cases} \tag{C.8}$$

and the initial total volume $|v_{0,ges}| = \sum_{\alpha \in \mathcal{PH}} |v_{0,\alpha}|$ reads

$$c_{1,\alpha} = c_{0,\alpha} - \sum_{\beta \in \mathcal{PH}, \beta \neq \alpha} \frac{|v_{0,\alpha}|}{|v_{0,ges}|^2} |\Delta v_\beta| + \sum_{\beta \in \mathcal{PH}, \beta \neq \alpha} \frac{|v_{0,\beta}|}{|v_{0,ges}|^2} |\Delta v_\alpha|. \tag{C.9}$$

Introducing the initial volume fraction $c_{0,\alpha} = |v_{0,\alpha}| / |v_{0,ges}|$, one obtains

$$c_{1,\alpha} = c_{0,\alpha} - \sum_{\beta \in \mathcal{PH}, \beta \neq \alpha} c_{0,\alpha} \frac{|\Delta v_\beta|}{|v_{0,ges}|} + \sum_{\beta \in \mathcal{PH}, \beta \neq \alpha} c_{0,\beta} \frac{|\Delta v_\alpha|}{|v_{0,ges}|}. \tag{C.10}$$

By enhancing fractures, the relation

$$\frac{|\Delta v_\alpha|}{|v_{0,ges}|} = c_{0,\alpha} \frac{|\Delta v_\alpha|}{|v_{0,\alpha}|} = c_{0,\alpha} \mathrm{tr}\left(\varepsilon_\alpha\right) \tag{C.11}$$

can be found leading to relation (C.1) and (C.3).

Appendix D

Partial Derivatives of the Free Energy

The partial derivatives of the macroscopic Helmholtz free energy are shown in this section.

The derivative of the Helmholtz free energy, with respect to the eigenstrain tensor of phase α, reads

$$\frac{\partial \bar{\psi}}{\partial \boldsymbol{\varepsilon}_{\alpha}^{E}} = -\frac{c_{\alpha}}{\bar{\rho}} \bar{\mathbb{C}}^{-1} \mathbb{C}_{\alpha} \mathbb{A}_{\alpha} \bar{\mathbb{C}} \left[\bar{\boldsymbol{\varepsilon}}^{E} \right] - \frac{c_{\alpha}}{2\bar{\rho}} \mathbb{C}_{\alpha} \left[\boldsymbol{A}_{\alpha}^{E} + \boldsymbol{\varepsilon}_{\alpha}^{E} \right] - \frac{c_{\alpha}}{2\bar{\rho}} \left(\frac{\partial \boldsymbol{A}_{\alpha}^{E}}{\partial \boldsymbol{\varepsilon}_{\alpha}^{E}} + \mathbb{I}^{S} \right) \mathbb{C}_{\alpha} \left[\boldsymbol{\varepsilon}_{\alpha}^{E} \right], \tag{D.1}$$

where the derivative of the nonlinear contribution of the localization relation, with respect to the eigenstrain tensor of phase α, is given by

$$\frac{\partial \boldsymbol{A}_{\alpha}}{\partial \boldsymbol{\varepsilon}_{\alpha}^{E}} = -\mathbb{R}_{\alpha} \mathbb{C}_{\alpha} + c_{\alpha} \mathbb{A}_{\alpha} \mathbb{R}_{\alpha} \mathbb{C}_{\alpha}. \tag{D.2}$$

The derivative of the Helmholtz free energy, with respect to the accumulated plastic strain of phase α, corresponds to the derivative of the hardening potential, what yields the Voce hardening law

$$\frac{\partial \bar{\psi}}{\partial \varepsilon_{\alpha}^{p}} = \frac{\partial \bar{\psi}^{i}}{\partial \varepsilon_{\alpha}^{p}} = c_{\alpha} \left(\Theta_{\alpha}^{\infty} \varepsilon_{\alpha}^{p} + \left(\sigma_{\alpha}^{\infty} - \sigma_{\alpha}^{0} \right) \left(1 - \exp\left(\frac{\Theta_{\alpha}^{\infty} - \Theta_{\alpha}^{0}}{\sigma_{\alpha}^{\infty} - \sigma_{\alpha}^{0}} \varepsilon_{\alpha}^{p} \right) \right) \right). \tag{D.3}$$

Neglecting the dependency of the stiffness tensors and the dependency of the inelastic effects of all phases $\mathbb{C}_\alpha$ on temperature, the partial derivative of the macroscopic Helmholtz free energy with respect to temperature reads

$$\frac{\partial \bar{\psi}}{\partial \theta} = -\frac{\bar{\alpha}}{\bar{\rho}} \text{tr}\left(\bar{\mathbb{C}}\left[\bar{\varepsilon} - \bar{\varepsilon}^E\right]\right) - \frac{\bar{\alpha}}{\bar{\rho}} \text{tr}\left(\bar{\mathbb{C}}\left[\bar{\varepsilon}^E\right]\right) + \frac{9}{\bar{\rho}} \left\langle K\alpha^2 \right\rangle \Delta\theta - \frac{1}{2\bar{\rho}} \left\langle \alpha \text{tr}\left(\mathbb{C}\left[\boldsymbol{A} + \varepsilon^E\right]\right) + \varepsilon^E \cdot \mathbb{C}\left[\frac{\partial \boldsymbol{A}}{\partial \theta} + \alpha \boldsymbol{I}\right] \right\rangle - \frac{1}{\bar{\rho}} \left\langle \kappa_c \right\rangle \left(\ln\left(\frac{\theta}{\theta_0}\right) - 1 \right), \quad \text{(D.4)}$$

where the derivative of the nonlinear contribution of the localization relation, with respect to temperature, is given by

$$\frac{\partial \boldsymbol{A}_\alpha}{\partial \theta} = -\mathbb{R}_\alpha \mathbb{C}_\alpha \left[\alpha_\alpha \boldsymbol{I}\right] + \mathbb{A}_\alpha \left\langle \mathbb{R}\mathbb{C}\left[\alpha \boldsymbol{I}\right] \right\rangle. \quad \text{(D.5)}$$

Appendix E

Material Parameter of 22MnB5 in Literature

In this appendix, measurement data of the steel 22MnB5 concerning the yield curves of the austenitic phase, the gap- and contact-pressure-dependence of the heat transfer coefficient, the temperature dependence of the friction coefficient, and the temperature dependence of the specific heat and thermal conductivity of the different phases are documented.

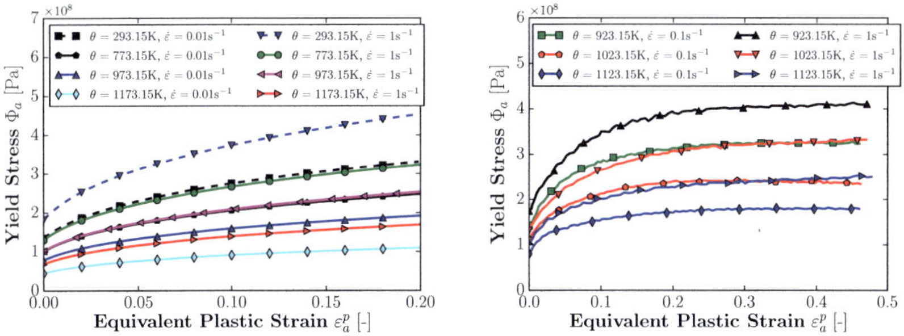

Figure E.1: Temperature- and strain-rate-dependent yield curves of austenite of 22MnB5 (left: Bok et al. (2015); right: Hochholdinger (2012)

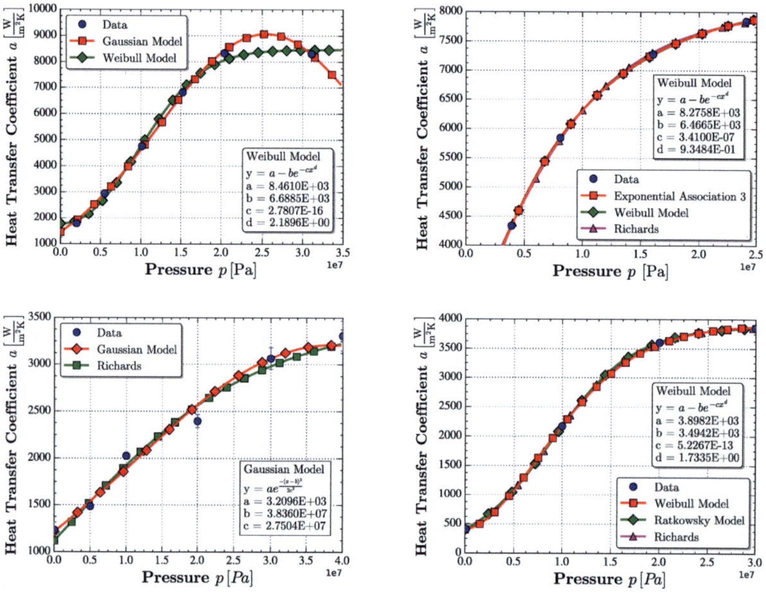

Figure E.2: Contact pressure dependence of the heat transfer coefficient (top left: Abdulhay et al. (2012); top right: Caron et al. (2014); bottom left: Bosetti et al. (2010); bottom right: Merklein et al. (2009))

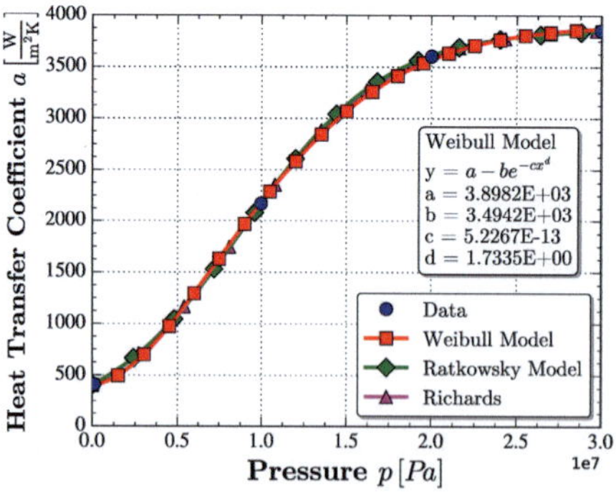

Figure E.3: Gap dependence of the heat transfer coefficient (Merklein et al., 2009)

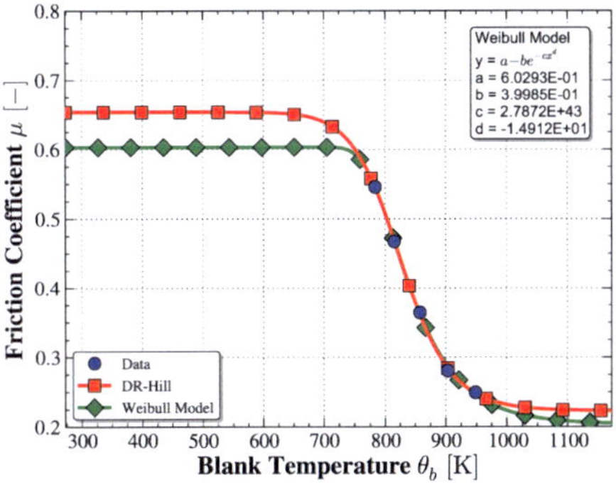

Figure E.4: Temperature dependence of the friction coefficient (Geiger et al., 2008)

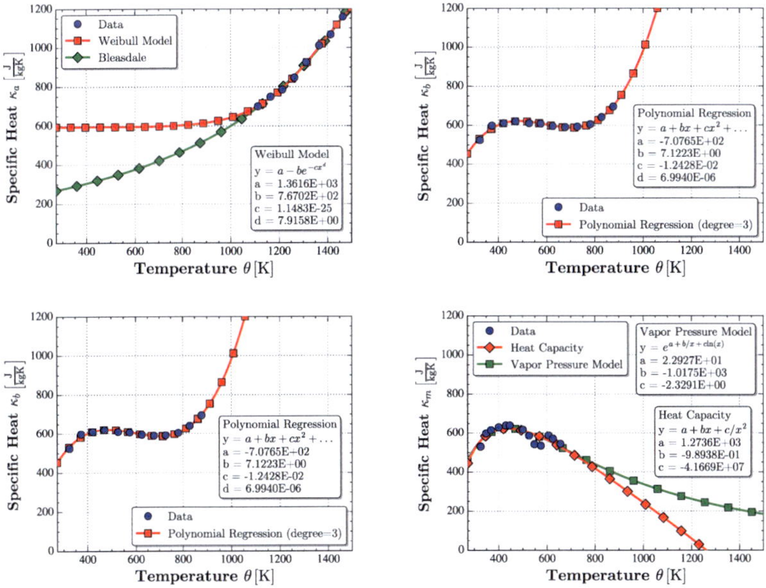

Figure E.5: Temperature dependence of the specific heat of the different phases (Tang et al., 2014) (top left: austenite; top right: ferrite; bottom left: bainite; bottom right: martensite)

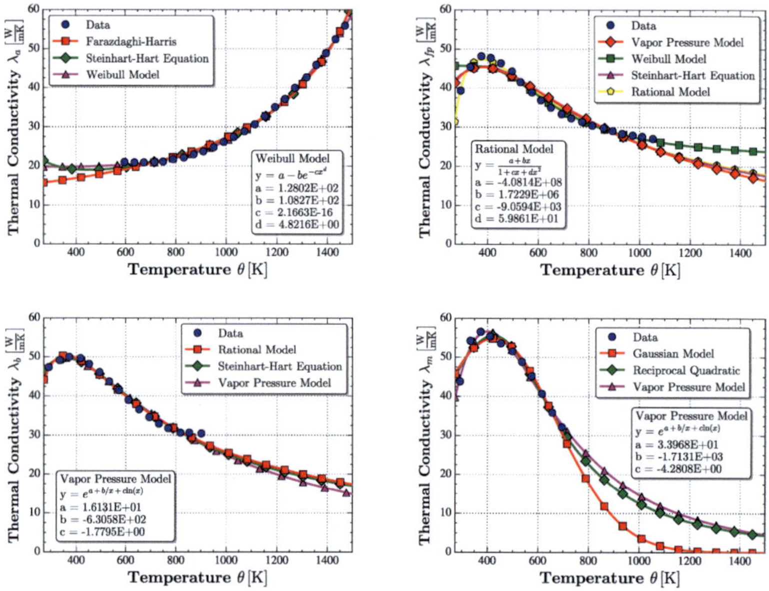

Figure E.6: Temperature dependence of the thermal conductivity of the different phases (Tang et al., 2014) (top left: austenite; top right: ferrite; bottom left: bainite; bottom right: martensite)

Appendix F

Experimental and Numerical Results of the U-Shape Demonstrator

The final shape and residual stresses of the simulation of the hot stamping process of u-shaped parts (see Figure F.2) are shown in this appendix.

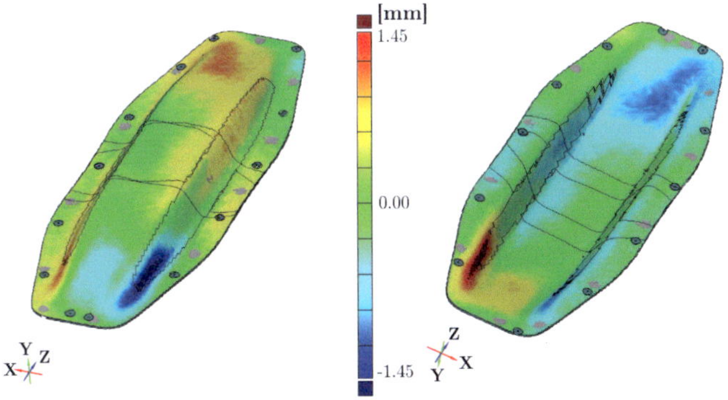

Figure F.1: Comparison of the final u-shaped parts of the process parameter sets 900-180-100-6 and 900-180-100-10

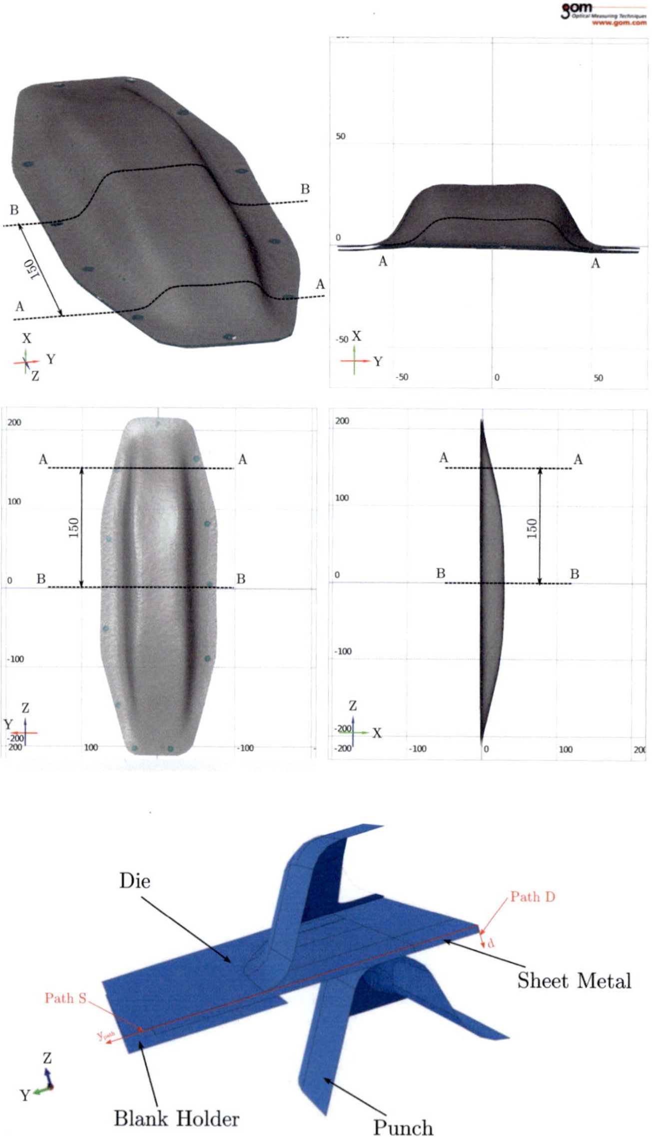

Figure F.2: Hot stamped u-shaped part of the process parameter set 900-180-100-6 (top and middle) and model for the simulation of the hot stamping process of the u-shape geometry (bottom)

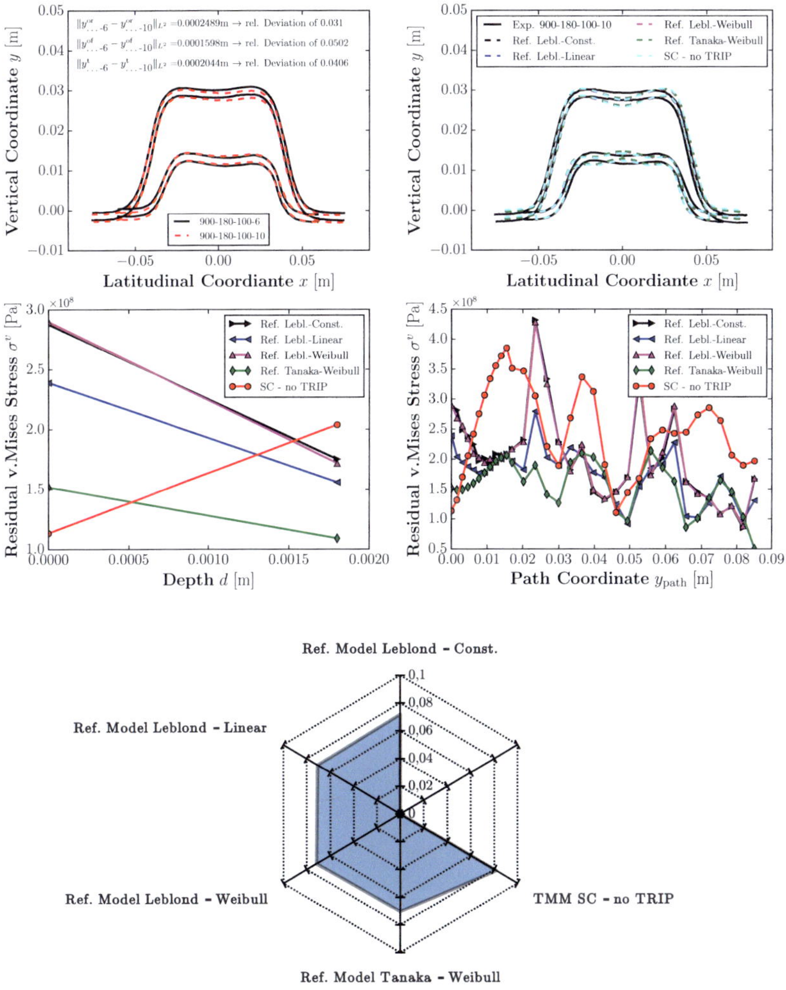

Figure F.3: Comparison of the final u-shape at the cut A-A and B-B of the process parameter sets 900-180-100-6 and 900-180-100-10 (top left), shape of the simulated u-shaped hot stamped part at the cut A-A and B-B of the reference model and the TMM model (top right), the v. Mises residual stresses at the Path D and at the Path S (middle left and right), and total relative deviation of the u-shaped geometry gained from experiments and simulation (bottom)

Frequently used acronyms, symbols, and operators

Latin Letters

A_α	Activation energy for the phase transformation of phase α
Ae_1	Equilibrium temperature, which is the lower limit of the regime at which ferrite and pearlite can coexist
Ae_3	Equilibrium temperature, which is the upper limit of the regime at which ferrite and austenite can coexist
Bs	Bainite formation start temperature
D	Total dissipation of a material
D^θ	Thermal dissipation of a material
$D^{m,p}$	Plastic part of the mechanical dissipation of a material
$D^{m,tr}$	TRIP part of the mechanical dissipation of a material
$D^{m,t}$	Mechanical dissipation of a material due to transformation strain
D^m	Mechanical dissipation of a material
E	Young's modulus
G	Shear modulus

K	Bulk modulus
Ms	Martensite formation start temperature
Q	Dissipated inelastic energy
R	Universal gas constant
S	ASTM austenite grain size number
W	Total inelastic energy
W^{θ}	Total thermal energy
$\boldsymbol{A}_{\alpha}$	Eigenstrain influence tensor of phase α
$\boldsymbol{D}$	Rate of deformation tensor
$\boldsymbol{F}$	Deformation gradient
$\boldsymbol{H}$	Displacement gradient
$\boldsymbol{I}$	Second-order identity tensor
$\boldsymbol{L}$	Velocity gradient
$\boldsymbol{L}^{\theta}$	Thermal Hill's constraint tensor
$\boldsymbol{R}$	Orthogonal orientation tensor
$\boldsymbol{U}$	Right Cauchy-Green tensor
$\boldsymbol{V}$	Left Cauchy-Green tensor
$\boldsymbol{W}$	Spin tensor
$\boldsymbol{X}$	Position vector to a material point in the reference placement with material coordinates
$\boldsymbol{f}$	Body force vector in the current placement
$\mathbb{A}_{\alpha}$	Fourth-order localization tensor of phase α
$\mathbb{C}$	Fourth-order stiffness tensors
$\mathbb{C}_0$	Stiffness tensor of the homogeneous comparison material
$\mathbb{I}^s$	Fourth-order identity tensor on symmetric second-order tensors
$\mathbb{L}$	Hill's constraint tensor
$\mathbb{P}^{1,iso}$	First fourth-order projector
$\mathbb{P}^{2,iso}$	Second fourth-order projector
$\mathbb{P}_0$	Fourth-order polarization tensor
$\boldsymbol{g}$	Temperature gradient

$\boldsymbol{p}$	Stress polarization tensor
$\boldsymbol{q}$	Heat flux vector through the body's surface in the current placement
$\boldsymbol{t}$	Stress vector on the body's surface in the current placement
$\boldsymbol{u}$	Displacement vector
$\boldsymbol{v}$	Velocity vector
$\boldsymbol{w}$	Relative velocity vector of a moving singularity through a body
$\boldsymbol{x}(t)$	Position vector to a material point in the current placement with spatial coordinates
$\mathcal{B}$	Set of all material points defining the body
$\mathcal{C}$	Set of the chemical components of the steel
$\mathcal{H}$	Heaviside function
$\mathcal{PH}$	Set of possible phases on the microstructure
$\mathcal{P}$	Material point
σ_α^v	V. Mises stress of phase α
$\tilde{c}_\alpha$	Normalized volume fraction of phase α
c_α^r	Remaining austenitic volume fraction before the transformation of phase α
c_α	Volume fraction of phase α
d_{L^2}	L^2-norm-induced metrics
$d_{L^2}^{\mathrm{r}el}$	Relative deviation based on the L^2-norm-induced metrics
e	Internal energy
k^{tr}	TRIP parameter
u	Uncertainty of a measurement with n repetitions
v	Set of material points of a body
w	Energy sources and sinks in the body
$\|v\|$	Volume of a set
$a, b, A, B, \ldots$	Scalar quantities
$\boldsymbol{a}, \boldsymbol{b}, \boldsymbol{c}, \ldots$	First-order tensors

$\boldsymbol{A}, \boldsymbol{B}, \boldsymbol{C}, \ldots$	Second-order tensors
$\mathbb{A}, \mathbb{B}, \mathbb{C}, \ldots$	Fourth-order tensors

Greek Letters

$\Delta\theta$	Difference of the absolute temperature and the reference temperature
α	Isotropic thermal expansion coefficient
β	Taylor-Quinney factor
χ_α	Indicator function of phase α
$\delta(\cdot)$	Dirac delta distribution
η	Entropy
$\boldsymbol{\alpha}$	Thermal expansion tensor
$\boldsymbol{\chi}_t$	Placement of a body at time t
$\boldsymbol{\varepsilon}$	Infinitesimal strain tensor
$\boldsymbol{\varepsilon}^E$	Eigenstrain tensor
$\boldsymbol{\varepsilon}^e$	Elastic part of the infinitesimal strain tensor
$\boldsymbol{\varepsilon}^i$	Inelastic part of the infinitesimal strain tensor
$\boldsymbol{\varepsilon}^p$	Plastic strain tensor
$\boldsymbol{\varepsilon}^{tr}$	TRIP strain tensor
$\boldsymbol{\varepsilon}^t$	Transformation strain tensor
$\boldsymbol{\lambda}$	Thermal conductivity tensor
$\boldsymbol{\lambda}_0$	Thermal conductivity tensor of the homogeneous comparison material
$\boldsymbol{\pi}$	Heat flux polarization
$\boldsymbol{\sigma}$	Cauchy stress tensor
γ^{tr}	Shear softening parameter due to TRIP
ι	Thermodynamical conjugate force of the internal variable
ι^p	Thermodynamical conjugate force of the accumulated plastic strain

κ	Specific heat capacity
κ^c	Constant part of the specific heat capacity
λ	Isotropic thermal conductivity
ν	Poisson's ratio
ϕ	Yield surface
ψ	Helmholtz free energy
ψ^i	Inelastic part of the Helmholtz free energy
ρ	Density in the current placement
σ^0	Initial yield stress
σ_α^c	Thermodynamical conjugate force of the volume fraction of phase α
τ_α	Spherical parameter of the HS localization tensor of phase α
θ	Absolute temperature
θ_0	Reference temperature
ε^p	Equivalent or accumulated plastic strain
φ	TRIP saturation function
ς_α	Deviatoric parameter of the HS localization tensor of phase α
ζ	Mean volume change due to transformation

Operators

$[\![\cdot]\!]$	Jump of a quantity at the grain boundary
$\langle\cdot\rangle$	Volume or ensemble average of a quantity
$(\cdot)'$	Deviatoric part of a second-order tensor
$(\cdot)^\circ$	Spheric part of a second-order tensor
$(\cdot)^{\mathrm{T}_H}$	Major transpose of a fourth-order tensor
$(\cdot)^{\mathrm{T}}$	Transpose of a second-order tensor
$[\cdot]$	Linear mapping of a second-order tensor, e.g., $\boldsymbol{A} = \mathbb{C}[\boldsymbol{B}]$

$\|\cdot\|$	Frobenius Norm of a tensorial quantity
$\bar{(\cdot)}$	Macroscopic quantity
$(\cdot)_\alpha$	Phase volume average of a quantity
$\operatorname{div}(\cdot)$	Eulerian divergence of a tensor
$\operatorname{Div}(\cdot)$	Lagrangian divergence of a tensor
$\operatorname{grad}(\cdot)$	Eulerian gradient of a tensor
$\operatorname{Grad}(\cdot)$	Lagrangian gradient of a tensor
$\cdot$	Scalar product between tensors, e.g., $a = \boldsymbol{A} \cdot \boldsymbol{B}$
$\otimes$	Dyadic product, e.g., $\mathbb{A} = \boldsymbol{A} \otimes \boldsymbol{B}$
$\square$	Special tensor product, e.g., $(\boldsymbol{A} \square \boldsymbol{B})\, \boldsymbol{C} = \boldsymbol{A}\boldsymbol{C}\boldsymbol{B}$
$\star$	Rayleigh product, e.g., $\boldsymbol{Q} \star \boldsymbol{A} = \boldsymbol{Q}\boldsymbol{A}\boldsymbol{Q}^{\mathrm{T}}$
$\operatorname{tr}(\cdot)$	Trace of a second-oder tensor
$\operatorname{sym}(\cdot)$	Symmetric part of a second-oder tensor
$\operatorname{skw}(\cdot)$	Skew-symmetric part of a second-oder tensor

Bibliography

ABAQUS, January 2016. DS Simulia. `http://www.3ds.com/products-services/simulia/products/abaqus/`.

Abdollahpoor, A., Chen, X., Pereira, M., Xiao, N., Rolfe, B., 2016. Sensitivity of the final properties of tailored hot stamping components to the process and material parameters. J. Mater. Process. Technol. 228, 125–136.

Abdulhay, B., Bourouga, B., Dessain, C., 2011. Experimental and theoretical study of thermal aspects of the hot stamping process. Appl. Therm. Eng. 31, 674–685.

Abdulhay, B., Bourouga, B., Dessain, C., 2012. Thermal contact resistance estimation: influence of the pressure contact and the coating layer during a hot forming process. Int. J. Mater. Form. 5, 183–197.

Achenbach, M., 1989. A model for an alloy with shape memory. Int. J. Plast. 5 (4), 371–395.

Ackerström, P., Bergman, P., Oldenburg, M., 2007. Numerical implementation of a constitutive model for simulation of hot stamping. Modell. Simul. Mater. Sci. Eng. 15, 105–119.

Ackerström, P., Oldenburg, M., 2006. Austenite decomposition during press hardening of a boron steel- computer simulation and test. J. Mater. Process. Technol. 174, 399–406.

Alexander, K., Becher, P., Wang, X., Hsueh, C., 1995. Internal stresses and the martensite start temperature in alumina-zirconia composites: effects of composition and microstructure. J. Am. Ceram. Soc. 78, 2, 291–296.

Ali, A., Bhadeshia, H., 1990. Nucleation of widmanstitten ferrite. J. Mater. Sci. Technol. 6, 781–784.

Andrews, K., 1965. Empirical formulae for the calculation of some transformation temperatures. J. Iron and Steel Inst. 203, 721–727.

Ashby, M., Easterling, K., 1984. The transformation hardening of steel surfaces by laser beams - I. hypo-eutectoid steels. Acta Metall. 32 (11), 1935–1948.

Avrami, M., 1939. Kinetics of phase change i. general theory. J. Chem. Phys. 7, 1103–1112.

Aziz, N., Aqida, S., 2013. Optimization of quenching process in hot press forming of 22MnB5 steel for high strength properties. IOP Conf. Ser. - Mater. Sci. Eng. 50, 1–6.

Bain, E., Dunkirk, N., 1924. The nature of martensite. Trans. AIME 70, 25.

Balzani, D., Brands, D., Schröder, J., 2014. Construction of statistically similar representative volume elements. Plast. Beyond 550, 355–412.

Bao, J., Liu, H., Xing, Z., Song, B., Yang, Y., 2013. Springback of hot stamping and die quenching with ultra-high-strength boron steel. Eng. Rev. 33, 3, 151–156.

Bariani, P., Bruschi, S., Ghiotti, A., Turetta, A., 2008. Testing formability in the hot stamping of hss. CIRP Ann. - Manuf. Technol. 57, 265–268.

Bartel, T., Hackl, K., 2009. A micromechanical model for martensitic phase-transformations in shape-memory alloys based on energy-relaxation. ZAMM 89 (10), 792–809.

Böck, N., Holzapfel, G., 2002. A large strain continuum and numerical model for transformation induced plasticity (TRIP). WCCM 5, 10 p.

Benveniste, Y., 1987. A new approach to the application of Mori-Tanaka's theory in composite materials. Mech. Mater. 6, 147–157.

Bertram, A., 2008. Elasticity and Plasticity of Large Deformations. Springer Verlag, Berlin, Heidelberg.

Bertram, A., Krawietz, A., 2012. On the introduction of thermoplasticity. Acta Mech. 223, 2257–2268.

Berveiller, M., Zaoui, A., 1979. An extension of the self-consistent scheme to plastically flowing polycristals. J. Mech. Phys. Solids 26, 325–344.

Bhadeshia, H., 1985. Diffusional formation of ferrite in iron and its alloys. Prog. Mater. Sci. 29, 321–386.

Böhlke, T., Neumann, R., Rieger, F., 2014. Two-scale modeling of grain size and phase transformation effects. Steel Res. Int. 85 (6), 1018–1034.

Billur, E., Altan, T., September 2015. Exploring the 3rd international conference on hot stamping technology, part i. `https://ercnsm.osu.edu/sites/ercnsm.osu.edu/files/uploads/S_hot/642-7.pdf`.

BMW AG, 2009. Interne Untersuchung. Laborbericht (080205463).

Bok, H., Kim, S., Suh, D., Barlat, F., Lee, M., 2015. Non-isothermal kinetics model to predict accurate phase transformation and hardness of 22MnB5 boron steel. Mater. Sci. Eng. A 626, 67–73.

Bok, H., Lee, M.-G., Pavlina, E., Barlat, F., Kim, H.-D., 2011. Comparative study of the prediction of microstructure and mechanical properties for a hot-stamped b-pillar reinforcing part. Int. J. Mech. Sci. 53, 744–752.

Bok, H.-H., Choi, J. W., Barlat, F., Suh, D. W., Lee, M.-G., 2014. Thermo-mechanical-metallurgical modeling for hot press forming in consideration of the prior austenite deformation effect. Int. J. Plast. 58, 154–183.

Borsetto, F., Ghiotti, A., Bruschi, S., 2009. Investigation of the high strength steel al-si coating during hot stamping operations. Key Eng. Mater. 410-411, 289–296.

Bos, C., Sietsma, J., 2007. A mixed-mode model for partitioning phase transformations. Scripta Mater. 57, 12, 1085–1088.

Bosetti, P., Brouschi, S., Stoehr, T., Lechler, J., Merklein, M., 2010. Interlaboratory comparison for heat transfer coefficient identification in hot stamping of high strength steels. Int. J. Mater. Form. 3, 1, 817–820.

Boyadjiev, I., Thomson, P., Lam, Y., 1996. Computation of the diffusional transformation of continuoisly colled austenite for predicting the coefficient of thermal expansion in the numerical analysis of thermal stress. ISIJ Int. 36 (11), 1413–1419.

Brinson, L., 2004. Stress-induced transformation behavior of a polycrystalline NiTi shape memory alloy: micro and macromechanical investigations via in situ optical microscopy. J. Mech. Phys. Solids 52 (7), 1549–1571.

Brünig, M., Driemeier, L., 2007. Numerical simulation of taylor impact tests. Int. J. Plast. 23, 1979–2003.

Buchelnikov, V., Bosko, S., 2003. The kinetics of phase transformations in ferromagnetic shape memory alloys Ni-Mn-Ga. J. Magn. Magn. Mater. 258-259, 497–499.

Buchheit, T., Wellman, G., Battaile, C., 2005. Investigating the limits of polycrystal plasticity modeling. Int. J. Plast. 21, 221–249.

Budiansky, B., 1965. On the elastic moduli of some heterogeneous materials. J. Mech. Phys. Solids 13, 223–227.

Cáceres, C., Poole, W., 2002. Hardness and flow strength in particulate metal matrix composites. Mater. Sci. Eng. A 332, 311–317.

Capdvila, C., Caballero, F., Garcia de Andres, C., 2003. Analysis of effect of alloying elements on martensite start temperature of steels. Mater. Sci. Technol. 19, 581–586.

Caron, E., Daun, K., Wells, M., 2014. Experimental heat transfer coefficient measurements during hot forming die quenching of boron steel at high temperatures. Int. J. Heat Mass Transfer 71, 396–404.

Casas, B., Latre, D., Rodriguez, N., Valls, I., 2008. Tailor made tool materials for the present and upcoming tooling solutions in hot sheet metal forming. 1st International Conference on Hot Sheet Metal Forming of High Performance Steels, Kassel, Germany 1, 23–35.

Caseiro, J., Oliveira, J., Andrade-Campos, A., 2011. Thermomechanical modelling strategies for multiphase steels. Int. J. Mech. Sci. 53, 720–733.

Chaboche, J., Kruch, S., Maire, J., Pottier, T., 2001. Towards a micromechanics based inelastic and damage modeling of composites. Int. J. Plast. 17, 411–439.

Cheng, X., Xiao, N., Li, D., Li, G., Sun, G., 2014. The finite element analysis of austenite decomposition during continuous cooling in 22MnB5 steel. Model. Simul. Mater. Sci. Eng. 22 (5), 16p.

Cherkaoui, M., Berveiller, M., Lemoine, X., 2000. Couplings between plasticity and martensitic phase transformation: overall behavior of polycrystalline TRIP steels. Int. J. Plast. 16 (10-11), 1215–1241.

Cherkaoui, M., Berveiller, M., Sabar, H., 1998. Micromechanical modeling of martensitic transformation induced plasticity (TRIP) in austenitic single crystals. Int. J. Plast. 14 (7), 597–626.

Cho, L., Kang, H., Lee, C., De Cooman, B., 2014. Microstructure of liquid metal embrittlement cracks on Zn-coated 22MnB5 press-hardened steel. Scr. Mater. 90-91, 25–28.

Christ, D., Reese, S., 2009. A finite element model for shape memory alloys considering thermomechanical couplings at large strains. Int. J. Solids Struct. 46 (20), 3694–3709.

Christian, J., 2002. The theory of transformations in metals and alloys. Elsevier Science, Oxford.

Cisse, C., Wael, Z., Zineb, T., 2016. A review of constitutive models and modeling techniques for shape memory alloys. Int. J. Plast. 76, 244–284.

Coleman, B., Noll, W., 1963. The thermodynamics of elastic materials with heat conduction and viscosity. Arch. Rational Mech. Anal. 13, 167–178.

Dahan, Y., Chastel, Y., Duroux, P., Hein, P., Massoni, E., 2006. Formability investigations for the hot stamping process. Int. Deep Drawing Res. Group (IDDRG), Int. Conf. , 8 p.

De Oliveira, W., Savi, M., Pachero, P., de Souza, L., 2010. Thermomechanical analysis of steel cylinders quenching using a constitutive model with diffusional and non-diffusional phase transformation. Mech. Mater. 42, 31–43.

Dederichs, P. H., Zeller, R., 1973. Variational treatment of the elastic constants of disordered materials. Z. Physik A 259, 103–116.

Denis, S., E., G., Simon, A., Beck, G., 1985. Stress-phase-interactions- basic principles, modelling, and calculation of internal stresses. Mater. Sci. Technol. 1, 805–814.

Desalos, Y., 1981. Comportement dilatométrique et mécanique de l'austénite métastable d'un acier A533. IRSID Rapport n° 95349401 .

Desalos, Y., Giusti, J., Gunsberg, F., 1982. Deformations et contraintes lors du traitement thermique de pieces en acier. Institut de Recherches de la Siderurgie Francaise RE 902.

DFG Forschergruppe 552, February 2016. Grundlagen der Warmblechumformung von höchstfesten Vergütungsstählen. http://www.lft.uni-erlangen.de/SEITEN/FG552/.

Diani, J., Berveiller, M., Sabar, H., 1993. Global behaviour related with interface movement; the case of transformation induced plasticity. Mater. Sci. Forum 123-125, 101–110.

Dixit, P., Dixit, U., 2008. Modeling of Metal Forming and Machining Processes. Springer.

Durrenberger, L., Wilsius, J., Hein, P., 2009. Impact of the constitutive model on the hot-forming numerical predictions of usibor 1500p. 2nd Int. Conf. on Hot Sheet Metal Forming of High-Performance Steel , 51–58.

Dvorak, G., 1992. Transformation field analysis of inelastic composite materials. Proc. Roy. Soc. A 437, 311–327.

Equihua, F., Salinas, A., 2010. Determination of equilibrium transformation temperatures Ae3 and Ae1 for low-carbon steels using the in situ high-temperature x-ray diffraction technique. Powder Diffraction 25 (1), 31–37.

Eshelby, J., 1957. The determination of the elastic field of an ellipsoidal inclusion and related problems. Proc. R. Soc. London A241, 376–396.

Eyre, D., G.W., M., 1999. A fast numerical scheme for computing the response of composites using grid refinement. Eur. Phys. J. Appl. Phys. 6, 41.

Faderl, J., Radlmayr, K., 2006. Ultraform und ultraform-phs innovation made by voestalpine. Tagungsband zum 1. Erlanger Workshop Warmblechumformung , 130–149.

Fan, D., De Cooman, B., 2012. State-of-the-knowledge on coating systems for hot stamped parts. Steel Res. Int. 83, 5, 412–433.

Fanfoni, M., Tomellini, M., 1998. The Johnson-Mehl-Avrami-Kolmogorov model: A brief review. Il Nuovo Cimento D 20 (7-8), 1171–1182.

Farren, W., Taylor, G., 1925. The heat developed during plastic extension of metals. Proc. Roy. Soc. 17, 422–451.

Feyel, F., 1999. Multiscale FE2 elastovisplastic analysis of composite structures. Comput. Mater. Sci. 16, 344–354.

Fischer, F. D., Oberaigner, E., Tanaka, K., Nishimura, F., 1992. Transformation-induced plasticity in triaxially loaded steel specimens subjected to a martensitic transformation. Eur. J. Mech. A-Solids 11, 233–244.

Fischer, F. D., Oberaigner, E., Tanaka, K., Nishimura, F., 1998. Transformation induced pasticity revised an updated formulation. Int. J. Solids Struct. 35 (18), 2209–2227.

Fischer, F. D., Sun, Q., Tanaka, K., 1996. Transformation-induced plasticity (TRIP). Appl. Mech. Rev. 49 (6), 317–364.

Flir Systems, Inc., February 2016. `http://www.flir.de/home/`.

Fotiu, P., Nemat-Nasser, S., 1996. Overall properties of elastic- viscoplastic periodic composites. Int. J. Plast. 12 (2), 163–190.

Fritzen, F., 2011. Microstructural modeling and computational homogenization of the physically linear and nonlinear constitutive behavior of micro-heterogeneous materials. Dissertation, Karlsruhe Institute of Technology KIT Scientific Publishing, Karlsruhe.

Ganghoffer, J., Simonsson, K., 1998. A micromechanical model of the martensitic transformation. Mech. Mater. 27 (3), 125–144.

Geers, M., Kouznetsova, V., Brekelmans, W., 2010. Multi-scale computational homogenization: trends and challenges. J. Comput. Appl. Math. 234, 2175–2182.

Geiger, M., Merklein, M., Hoff, C., 2005. Basic investigations on the hot stamping steel 22MnB5. Adv. Mater. Res. 6-8, 795–804.

Geiger, M., Merklein, M., Lechler, J., 2008. Determination of tribological conditions within hot stamping. Prod. Eng. Res. Dev. 2, 269–276.

George, R., Bardelcik, A., Worswick, M., 2012. Hot forming of boron steels using heated and cooled tooling for tailored properties. J. Mater. Process. Technol. 212, 2386–2399.

Goedicke, S., Sepeur, S., Frenzer, G., Breyer, C., 2008. Wet chemical coating materials for hot sheet forming-anti scaling and corrosion protection. 1st International Conference on Hot Sheet Metal Forming of High-Performance Steel , 37–44.

GOM GmbH, February 2016. ATOS - Industrial 3D Scanning Technology. `http://www.gom.com/de/messsysteme/3d-scanner.html`.

Govindjee, S., Hall, G., 2000. A computational model for shape memory alloys. Int. J. Solids Struct. 37 (5), 735–760.

Greenwood, G., Johnson, R., 1965. The deformation of metals under small stresses during phase transformations. Proc. Roy. Soc. A, Math. Phys. Sci. 283, 403–422.

Greninger, A., Troiano, A., 1949. The mechanism of martensite formation. Trans. AIME 185, 590–598.

Gárlipp, W., Cilense, M., Novaes Gomes, S., 2001. Austenite decomposition of C-Mn steel containing boron by continuous cooling. J. Mater. Process. Technol. 114, 71–74.

Gross, D., Seelig, T., 2002. Bruchmechanik. Springer.

Guo, Y., 2011. Critical cooling rate and design of mould cooling system for hot forming of high-strength sheet. Adv. Mater. Res. 295-297, 1474–1478.

Guo, Z., 2004. On coherent transformations in steel. Acta Mater. 52, 5511–5518.

Hallberg, H., Hakansson, P., Ristinmaa, M., 2007. A constitutive model for the formation of martensite in austenitic steels under large strain plasticity. Int. J. Plast. 23, 1213–1239.

Hallberg, H., Hakansson, P., Ristinmaa, M., 2010. Thermo-mechanically coupled model of diffusionless phase transformation in austenitic steel. Int. J. Solids and Struct. 47, 1580–1591.

Hashin, Z., Shtrikman, S., 1962a. A variational approach to the theory of the elastic behaviour of polycrystals. J. Mech. Phys. Solids 10, 343–352.

Hashin, Z., Shtrikman, S., 1962b. A variational approach to the theory of the elastic behaviour of polycrystals. J. Mech. Phys. Solids 10, 343–352.

Haupt, P., 2002. Continuum Mechanics and Theory of Materials. Springer Verlag, Berlin, Heidelberg.

He, L., Zhao, G., Li, H., 2010. Measurement and analysis of time-temperature-transformation curves of boron steel 22MnB5. Appl. Mech. Mater. 29-32, 484–489.

Helm, D., Haupt, P., 2003. Shape memory behaviour: modelling within continuum thermomechanics. Int. J. Solids Struct. 40 (4), 827–849.

Hill, R., 1952. The elastic behaviour of a crystalline aggregate. Proc. Phys. Soc. A 65, 349–354.

Hill, R., 1963. Elastic properties of reinforced solids: some theoretical principles. J. Mech. Phys. Solids 11 (5), 357–372.

Hill, R., 1965a. Continuum micromechanics of elastoplastic polycrystals. J. Mech. Phys. Solids 13, 89–101.

Hill, R., 1965b. A self-consistent mechanics of composite materials. J. Mech. Phys. Solids 13, 213–222.

Hillert, M., 1957. The role of interfacial energy during solid-state phase transformation. Jernkontorets Analler. 141, 757–789.

Hillert, M., 1975. Diffusion and interface reactions in alloys. Metall. Trans. A 6, 5–19.

Hippchen, P., Lipp, A., Grass, H., Craighero, P., Fleischer, M., Merklein, M., 2016. Modelling kinetics of phase transformation for the indirect hot stamping process to focus on car body parts with tailored properties. J. Mater. Process. Technol. 228, 59–67.

Hömberg, D., 1995. A mathematical model for the phase transitions in eutectoid carbon steel. IMA J. Appl. Math. 54, 31–57.

Hömberg, D., 1996. A numerical simulation of the Jominey end-quench test. Acta Metall. 44, 4375–4385.

Hochholdinger, B., 2012. Simulation des Presshärteprozesses und Vorhersage der mechanischen Bauteileigenschaften nach dem Härten. Dissertation, ETH Zürich .

Hochholdinger, B., Hora, P., Grass, H., Lipp, A., 2011. Simulation of the press hardening process and prediction of the final mechanical material properties. Numisheet 2011, Am. Inst. Phys. , 618–625.

Hockett, J., Sherby, O., 1975. Large strain deformation of polycrystalline metals at low homologous temperatures. J. Mech. Phys. Solids 23 (2), 87–98.

Hoff, N., 1954. Approximate analysis of structures in the presence of moderately large creep deformations. Q. Appl. Math. 12, 49–55.

Holzapfel, G. A., 2000. Nonlinear Solid Mechanics. John Wiley and Sons Ltd., Sussex.

Hougardy, H., Yamazaki, K., 1986. An improved calculation of the transformation of steels. Steel Res. Int. 57 (9), 466–471.

Hu, P., Shi, D., Ying, L., Shen, G., Liu, W., 2015. The finite element analysis of ductile damage during hot stamping of 22MnB5 steel. Mater. Design 69, 141–152.

Husson, J., Dubois, F., Sauvat, N., 2011. A finite element model for shape memory behavior. Mech. Time-Depend. Mater. 15 (3), 213–237.

Hutchinson, J., 1976. Bounds and self-consistent estimates for creep and polycrystalline materials. Proc. Roy. Soc. A 348, 101–127.

Inoue, T., Wang, Z., 1985. Coupling between stress, temperature, and metallic structures during processes involving phase transformations. Mater. Sci. Technol. (1), 845–850.

Itskov, M., 2013. Tensor Algebra and Tensor Analysis for Engineers. Springer Verlag, Berlin, Heidelberg.

Jani, J., Leary, M., Subic, A., Gibson, M., 2014. A review of shape memory alloy research, applications and opportunities. Mater. and Design 56, 1078–1113.

Jöchen, K., 2013. Homogenization of the linear and non-linear mechanical behavior of polycrystals. Dissertation, Karlsruhe Institute of Technology KIT Scientific Publishing, Karlsruhe.

Jöchen, K., Böhlke, T., 2012. Prediction of texture evolution in rolled sheet metals by using homogenization schemes. Key Eng. Mater. 504-506, 649–654.

Johnson, G., Cook, W., 1983. A constitutive mode and data for metals subjected to large strains, high strain-rates and high temperature. 7th Symp. on Ballistic , 541–547.

Johnson, W., Mehl, R., 1939. Reaction kinetics in processes of nucleation and growth. Trans. Am. Inst. Min. Metall. Eng. 135, 414–458.

Jones, S., Bhadeshia, H., 1997. Kinetics of the simultaneous decomposition of austenite into several transformation products. Acta Mater. 45,7, 2911–2920.

Jun, H., Kang, J., Seo, D., Kang, K., Park, C., 2006. Effects of deformation and boron on microstructure and continuous cooling transformation in low carbon hsla steels. Mater. Sci. Eng. A 422, 157–162.

Kamamoto, S., Nishimori, T., Kinoshita, S., 1985. Analysis of residual stress and distortion resulting from quenching in large low-alloy steel shafts. Mater. Sci. Technol. 1, 789–804.

Kanouté, P., Boso, D., Chaboche, J., Schrefler, B., 2009. Multiscale methods for composites: a review. Arch. Comput. Meth. Eng. 16 (1), 31–75.

Kapoor, R., Nemat-Nasser, S., 1998. Determination of temperature rise during high strain rate deformation. Mech. Mater. 27, 1, 1–12.

Karbasian, H., 2010. Formgenauigkeit und mechanische Eigenschaften pressgehärteter Bauteile. Dissertation, TU Dortmund, Shaker Verlag .

Karbasian, H., Tekkaya, A., 2010. A review on hot stamping. J. Mater. Process. Technol. 15, 2103–2118.

Kastner, O., Eggeler, G., Weiss, W., Ackland, G. J., 2011. Molecular dynamics simulation study of microstructure evolution during cyclic martensitic transformations. J. Mech. Phys. Solids 59 (9), 1888–1908.

Khan, A., Liu, H., 2012. Variable strain rate sensitivity in an aluminum alloy: response and constitutive modeling. Int. J. Plast. 36, 1–14.

Kim, H., Park, J., Lee, M., 2012. Finite element simulation of hot stamping of tubular beam and its application to die channel design. Mater. Trans. 53 (5), 838–846.

Kim, H.-K., Lee, S., Choi, H., 2015. Evaluation of contact heat transfer coefficient and phase transformation during hot stamping of a hat-type part. Mater. 8, 2030–2042.

Kim, N.-H., 2015. Introduction to Nonlinear Finite Element Analysis. Springer.

Kinsey, B., Liu, Z., Cao, J., 2000. A novel forming technology for tailor-welded blanks. J. Mater. Process. Technol. 99, 145–153.

Kirkaldy, J., Venugopalan, D., 1983. Prediction of microstructure and hardenability in low alloy steels. Int. Conerence Phase Transformation Ferrous Alloys , 125–148.

Klein, M., 2007. Einführung in die DIN-Normen. Vol. 14. Teubner B.G. GmbH, Stuttgart, Leipzig, Wiesbaden, hrsg. DIN Deutsches Institut für Normung e.V.

Koistinen, D., Marburger, R., 1959. A general equation prescribing the extent of the austenite-martensite transformation in pure iron-carbon alloys and plain carbon steels. Acta Metall. , 59–60.

Kollecka, R., Veita, R., Merklein, M., Lechler, J., Geiger, M., 2009. Investigation on induction heating for hot stamping of boron alloyed steels. CIRP Ann. - Manuf. Technol. 58, 1, 275–278.

Kolmogorov, A., 1937. Statistical theory of crystallization of metals. (in russian). Ivz. Akad. Nauk SSSR Ser. Mat. 3, 355–360.

Kondratiuk, J., Kuhn, P., 2011. Tribological investigation on friction and wear behaviour of coatings for hot sheet metal forming. Wear 270, 839–849.

Kondratiuk, J., Kuhn, P., Labrenz, E., Bischoff, C., 2011. Zinc coatings for hot sheet metal forming: comparison of phase evolution and microstructure during heat treatment. Surf. Coat. Technol. 205, 4141–4153.

Kooi, B., 2006. Extension of the Johnson-Mehl-Avrami-Kolmogorov theory incorporating anisotropic growth studied by Monte Carlo simulations. Phys. Rev. B 73, 054103.

Kouznetsova, V., Geers, M., 2008. A multi-scale model of martensitic transformation plasticity. Mech. Mater. 40 (8), 641–657.

Krüger, P., 1993. On the relation between non-isothermal and isothermal Kolmogorov-Johnson-Mehl-Avrami crystallization kinetics. J. Phys. Chem. Solids 54, 11, 1549–1555.

Kröner, E., 1958. Berechnung der elastischen Konstanten des Vielkristalls aus den Konstanten des Einkristalls. Z. Physik 151, 504–518.

Kröner, E., Sauthoff, G., 1972. Statistical Continuum Mechanics. Springer-Verlag, Wien.

Kubler, R., Berveiller, M., Buessler, P., 2011. Semi phenomenological modelling of the behavior of TRIP steels. Int. J. Plast. 27, 299–327.

Kurdjumov, G., Sachs, G., 1930. Über den Mechanismus der Stahlhärtung. Z. Physik 64, 325–332.

Labudde, T., Bleck, W., 2009. Formability characterization of press hardened steels. 2nd International Conference on Hot Sheet Metal Forming of High Performance Steels , 127–135.

Lamprecht, K., Deinzer, G., 2008. Hot sheet metal forming in automotive production. Proc. 2nd IWOTE , 145–154.

Leblond, J., 1989. Mathematical modelling of transformation plasticity in steels II: coupling with strain hardening phenomena. Int. J. Plast. 5, 573–591.

Leblond, J., Devaux, J., Devaux, J., 1989. Mathematical modelling of transformation plasticity in steels I: case of ideal-plastic phases. Int. J. Plast. 5, 551–572.

Leblond, J., Mottet, B., Devaux, J., 1986. A theoretical and numerical approach to the plastic behaviour of steels during phase transformations. i. derivation of general relations. J. Mech. Phys. Solids 34, 395–409.

Lechler, J., 2009. Grundlegende Untersuchungen zur Beschreibung und Modellierung des Werkstoffverhaltens von presshärtbaren Bor-Manganstählen. Dissertation, LFT Erlangen .

Lechler, J., Stöhr, T., Kuppert, A., Merklein, M., 2010. Basic investigations on hot stamping of tailor welded blanks regarding the manufacturing of lightweight components with functionally optimized mechanical properties. Trans. North Am. Manuf. Res. Inst. SME 38, 593–600.

Lee, M.-G., Kim, S.-J., Han, H., Jeong, W., 2009. Implicit finite element formulations for multi-phase transformation in high carbon steel. Int. J. Plast. 25, 1726–1758.

Lee, S.-J., Lee, Y., 2008. Finite element simulation of quench distortion in a low aloy steel incorporating transformation kinetics. Acta Mater. 56, 1482–1490.

Lee, S.-J., Pavlina, E., Van Tyne, C., 2010. Kinetics modeling of austenite decomposition for an end-quenched 1045 steel. Mater. Sci. Eng. A 527, 3186–3194.

Lemaitre, J., 1985. A continuous damage mechanics model for ductile fracture. J. Eng. Mater. Technol. 107, 83–89.

Lenze, F., Sikora, S., 2006. Herstellung von Karosseriebauteilen aus warmgeformten höchstfesten Stahlwerkstoffen. Tagungsbd. zum 1. Erlanger Workshop Warmblechumformung , 1–12.

Leroy, Y., Ponte Castañeda, P., 2001. Bounds of the self-consistent approximation for nonlinear media and implications for the second order method. C. R. Acad. Sci. Paris IIB 329 (8), 1203–1227.

Levin, V., 1967. Determination of the thermoelastic constants of composite materials. Trans. Soviet Acad. Sci. (Mech. of Solids) 6, 137–145.

Levitas, V., Preston, D., 2002. Three-dimensional Landau theory for multivariant stress-induced martensitic phase transformations. i. austenite-martensite. Phys. Rev. B 66 (134206), 1–9.

Li, M., Niebuhr, D., Meekisho, L., Atteridge, D., 1998. A computational model for the prediction of steel hardenability. Metall. Mater. Tans. B 29, 661–672.

Liang, W., Liu, Y., Zhu, B., Zhou, M., Zhang, Y., 2015. Conduction heating of boron alloyed steel in application for hot stamping. Int. J. Prec. Eng. Manuf. 16, 9, 1983–1992.

Liang, W., Wang, L., Liu, Y., Wang, Y., Zhang, Y., 2014. Hot stamping parts with tailored properties by local resistance heating. Procedia Eng. 81, 1731–1736.

Lin, J., Mohamed, M., Balint, D., Dean, T., 2013. The development of continuum damage mechanics-based theories for predicting forming limit diagrams for hot stamping applications. Int. J. Damage Mech. , 1–18.

Lin, T., 1957. Analysis of elastic and plastic strains of a fcc crystal. J. Mech. Phys. Solids 5, 143–149.

Liu, F., Sommer, F., Bos, C., Mittemeijer, E., 2007. Analysis of solid state phase transformation kinetics: models and recipes. Int. Mater. Rev. 52, 193–212.

Liu, H., Lei, C., 2015. Local heating-aided hot blanking of quenched ultra-high-strength steel br1500hs. Int. J. Adv. Manuf. Technol. 77, 1, 629–641.

Liu, H., Lei, C., Xing, Z., 2013. Cooling system of hot stamping of quenchable steel BR1500HS: optimization and manufacturing methods. Int. J. Adv. Manuf. Technol. 69, 211–223.

Liu, I., Cipolatti, R., Rincon, M., 2006. Incremental linear approximation for finite elasticity. ICNAAM Extended Abstracts , 1–4.

Lobo, P., Almeida, J., Guerreiro, L., 2015. Shape memory alloys behaviour: a review. Procedia Eng. 114, 776–783.

LS-Dyna, January 2016. Livermore Software Technology Corporation. `http://www.lstc.com/`.

Lusk, M., Jou, H., 1997. On the rule of additivity in phase transformation kinetics. Metall. Mater. Trans. A 28, 287–291.

Magee, C., 1969. The nucleation of martrensite. ASM Int. , 115–156.

Mahnken, R., Schneidt, A., Antretter, T., Ehlenbröker, U., Wolff, M., 2015. Multi-scale modeling of bainitic phase transformation in multi-variant polycrystalline low alloy steels. Int. J. Solids and Struct. 54, 156–171.

Mahnken, R., Wilmanns, S., 2011. A projected newton algorithm for simulation of multi-variant textured polycrystalline shape memory alloys. Comput. Mater. Sci. 50 (9), 2535–2548.

Mahnken, R., Wolff, M., Schneidt, A., Böhm, M., 2012. Multi-phase transformations at large strains - thermodynamic framework and simulation. Int. J. Plast. 39, 1–26.

Mandel, J., 1965. Une generalisation de la theorie de la plasticite de w.t. koiter. Int. J. Solids Struct. 1, 273–295.

Masson, R., Bornert, M., Suquet, P., Zaoui, A., 2000. An affine formulation for the prediction of the effective properties of nonlinear composites and polycrystals. J. Mech. Phys. Solids 48, 1203–1227.

Masson, R., Zaoui, A., 1999. Self-consistent estimates for the rate-dependent elastoplastic behaviour of polycrystalline materials. J. Mech. Phys. Solids 47, 1543–1568.

Materials Views, February 2016. Lightweight Construction with no Aluminum or Carbon. http://www.materialsviews.com/lightweight-construction-with-no-aluminum-or-carbon/.

Maynier, P., Jungmann, B., Dollet, J., 1978. Creusot-loire system for the prediction of the mechanical properties of low alloy steel products. Trans. Metall. Soc. AIME , 518–545.

MBW 1500, February 2016. Thyssenkrupp Steel. https://www.thyssenkrupp-steel.com/media/content_1/publikationen/pi_mb-w_de_nov_2014.pdf.

Merklein, M., Lechler, J., 2006. Investigation of the thermo-mechanical properties of hot stamping steels. J. Mater. Process. Technol. 177, 452–455.

Merklein, M., Lechler, J., Stoehr, T., 2009. Investigations on the thermal behavior of ultra high strength boron manganese steels within hot stamping. Int. J. Mater. Form. 2, 1, 259–262.

Merklein, M., Wieland, M., Lechner, M., Bruschi, S., Ghiotti, A., 2016. Hot stamping of boron steel sheets with tailored properties: A review. J. Mater. Process. Technol. 228, 11–24.

Michel, J., Moulinec, H., Suquet, P., 2000. A computational method based on augmented lagrangians and fast fourier trans- forms for composites with high contrast. Comput. Model. Eng. Sci. 1 (2), 79–88.

Michel, J., P., S., 2003. Nonuniform transformation field analysis. Int. J. Solids. Struct. 40, 6937–6955.

Michel, J., P., S., 2004. Computational analysis of nonlinear composite structures using the nonuniform transformation field analysis. Comput. Method. Appl. Mech. Eng. 193, 5477–5502.

Michelitsch, T., Mehnen, J., 2006. Evolutionäres Optimieren von Temperierbohrungssystemen. Tagungsband zum 1. Erlanger Workshop Warmblechumformung , 118–129.

Miehe, C., Schröder, J., Schotte, J., 1999. Computational homogenization analysis in finite plasticity. simulation of texture development in polycrystalline materials. Comput. Meth. Appl. Mech. Eng. 171, 317–418.

Min, J., Lin, J., Min, Y., 2013. Effect of thermo-mechanical process on the microstructure and secondary-deformation behavior of 22MnB5 steels. J. Mater. Process. Technol. 213 (6), 818–825.

Min, J., Lin, J., Min, Y., Li, F., 2012. On the ferrite and bainite transformation in isothermally deformed 22MnB5 steels. Mater. Sci. Eng. A 550 (6), 375–387.

Miokovic, T., 2005. Analyse des Umwandlungsverhaltens bei ein- und mehrfacher Kurzzeithaertung bzw. Laserstrahlhaertung des Stahls 42CrMo4. Dissertation, TH Karlsruhe Shaker Verlag.

Mitter, W., 1987. Umwandlungsplastizität und ihre Berücksichtigung bei der Berechnung von Eigenspannungen. Bornträger-Verlag, Stuttgart .

Molinari, A., Canova, G., Ahzi, S., 1987. A self-consistent approach of the large deformation polycristal viscoplasticity. Acta Metall. 35, 2983–2994.

Mori, K., Maeno, T., Yamada, H., Matsumoto, H., 2015. S1-Shot hot stamping of ultra-high strength steel parts consisting of resistance heating, forming, shearing and die quenching. Int. J. Machine Tools and Manuf. 89, 124–131.

Mori, K., Maki, S., Tanaka, Y., 2005. Warm and hot stamping of ultra high tensile strength steel sheets using resistance heating. CIRP Ann. - Manuf. Technol. 54, 1, 209–212.

Mori, K.-i., 2012. Smart hot stamping of ultra-high strength steel parts. Trans. Nonferrous Met. Soc. China 22, 496–503.

Mori, K.-i., Maeno, T., Mongkolkaji, K., 2013. Tailored die quenching of steel parts having strength distribution using bypass resistance heating in hot stamping. J. Mater. Process. Technol. 213, 508–514.

Moulinec, H., Suquet, P., 1998. A numerical method for computing the overall response of non-linear composites with complex microstructure. Comput. Meth. Appl. Mech. Eng. 157, 69–94.

Moumni, Z., Roger, F., Trinh, N., 2011. Theoretical and numerical modeling of the thermomechanical and metallurgical behavior of steel. Int. J. Plast. 27 (3), 414–439.

Naderi, M., 2007. Hot stamping of ultra high strength steels. Dissertation, RWTH Aachen .

Nemat-Nasser, S., 1999. Experimentally-based micromechanical modeling of metal plasticity with homogenization form micro-to-macro-scale properties. IUTAM Symp. on Micro- and Macrostr. Asprects of Thermoplast. , 101–113.

Neugebauer, R., Schieck, F., Polster, S., Mosel, A., Rautenstrauch, A., Schönherr, J., Pierschel, N., 2012. Press hardening - an innovative and challenging technology. Arch. Civil Mech. Eng. 12 (2), 113–118.

Neumann, R., Böhlke, T., 2013. Nonlinear homogenization of microstructures in steel with temperature-controlled phase transformation. PAMM 13 (1), 267–268.

Neumann, R., Böhlke, T., 2016. Hashin-Shtrikman type mean field model for the two-scale simulation of the thermomechanical processing of steel. Int. J. Plast. 77, 1–29.

Neumann, R., Schuster, S., Böhlke, T., Gibmeier, J., 2017. Two-scale simulation of the hot stamping process based on a Hashin-Shtrikman type mean field model. submitted to J. Mater. Process. Technol. .

Nikravesh, M., Naderi, M., Akbari, G., 2012. Influence of hot plastic deformation and cooling rate on martensite and bainite start temperatures in 22MnB5 steel. Mater. Sci. Eng. A 540, 24–29.

Nishiyama, Z., 1934. X-ray investigation on the mechanism of the transformation from face-centered cubic lattice to body-centered cubic lattice. Sci. Rep. Tohoku Imp. Univ. 23, 637–664.

Nolze, G., 2004. Characterization of the fcc/bcc orientation relationship by EBSD using pole figures and variants. Z. Metallkd. 95 (9), 744–755.

Nußkern, P., 2013. Prozesssimulation der Herstellung einsatzgehärteter pulvermetallurgischer Bauteile. Dissertation, Karlsruhe Institute of Technology KIT Scientific Publishing.

Oberste-Brandenburg, C., Bruhns, O., 2004. A tensorial description of the transformation kinetics of the martensitic phase transformation. Int. J. Plast. 20 (12), 2083–2109.

Offerman, S., van Dijk, N., Sietsma, J., Grigull, S., Lauridsen, E., Margulies, L., Poulsen, H., Rekveldt, M., van der Zwaag, S., 2002. Grain nucleation and growth during phase transformations. Sci. 298, 1003–1005.

Olsson, T., 2009. An LS-DYNA material model for simulations of hot stamping processes of ultra high strength steels. Proc. 7th European LS-DYNA Users Conference .

Ostoja-Starzewski, M., 2005. Towards stochastic continuum thermodynamics. J. Non-Equilib. Thermodyn. 27, 335–348.

Ostwald, R., 2015. Modelling and simulation of phase-transformations in elasto-plastic polycrystals. Koffler Druck Management GmbH, Dortmund.

Ostwald, R., Bartel, T., Menzel, A., 2010. A computational micro-sphere model applied to the simulation of phase-transformations. ZAMM - J. Appl. Math. Mech. 90 (7-8), 605–622.

Ostwald, R., Bartel, T., Menzel, A., 2012. Phase-transformations interacting with plasticity - a micro-sphere model applied to TRIP steel. Comput. Mater. Sci. 64, 12–16.

Pandit, A., Bhadeshia, H., 2011. Mixed diffusion-contolled growth of pearlite in binary steel. Proc. Roy. Soc. A 467, 508–521.

Panico, M., Brinson, L., 2007. A three-dimensional phenomenological model for martensite reorientation in shape memory alloys. J. Mech. Phys. Solids 55 (11), 2491–2511.

Papatriantafillou, I., Agoras, M., Aravas, N., Haidemenopoulos, G., 2006. Constitutive modeling and finite element methods for TRIP steels. Comput. Meth. Appl. Mech. Eng. 195, 5094–5114.

Patoor, E., Eberhardt, A., Berveiller, M., 1988. Thermomechanical behaviour of shape memory alloys. Arch. Mech. 40, 775–794.

Perzyna, P., 1971. Terhmodynamic theory of viscoplasticity. Adv. Appl. Mech. 9, 315–354.

phs-ultraform, February 2016. voestalpine steel division. `https://www.voestalpine.com/division_stahl/content/download/3993/30758/file/Folder_phs-ultraform_E_16052013_ES.pdf`.

phs-ultraform 1500 Z140, February 2016. voestalpine steel division. `https://www.voestalpine.com/division_stahl/content/download/3987/30705/file/DB_phs-ultraform_1500Z140_D_12032013.pdf`.

Pitsch, W., 1967. Archiv Eisenhüttenwesen 38, 853.

Pitsch, W., Sauthoff, G., 1992. Effect of substitutuina alloying elements; Steel: a Handbook for Mater. Res. and Eng. Springer Verlag, Berlin.

Ponte Castañeda, P., 1992. New variational principles in plasticity and their application to composite materials. J. Mech. Phys. Solids 40, 1757–1788.

Ponte Castañeda, P., 2002. Second-order homogenisation estimates for nonlinear composites incorporating field fluctuations: i - theory. J. Mech. Phys. Solids 50 (4), 737–757.

Ponte Castañeda, P., Suquet, P., 1998. Nonlinear composites. Adv. Appl. Mech. 34, 171–302.

Ponte Castañeda, P., Suquet, P., 1998. Nonlinear composites. Advances in Applied Mechanics 34, 171–302.

Pradell, T., Crespo, D., Clavaguera, N., Clavaguera-Mora, M., 1998. Diffusion contolled grain growth in primary crystallization: Avrami exponents revisited. J. Phys. Condens. Matter 10, 3833–3844.

Renard, J., 1987. Etude de l'initiation de l'endommagement dans la matrice d'un matériau composite par une méthode d'homogénisation. Aerosp. Sci. Technol. 6, 37–51.

Rendler, N., Vigness, I., 1966. Hole-drilling strain-gage method of measuring residual stresses. Exp. Mech. 6 (12), 577–586.

Reuss, A., 1929. Berechnung der Fliessgrenze von Mischkristallen auf Grund der Plastizitätbedingung für Einkristalle. Z. Angew. Math. Mech. 9, 49–58.

Rios, P., 2005. Relationship between non-isothermal transformation curves and isothermal and non-isothermal kinetics. Acta Mater. 53, 4893–4901.

Rosakis, P., Rosakis, A., Ravichandran, G., Hodowany, J., 2000. A thermodynamic internal variable model for the partition of plastic work into heat and stored energy in metals. J. Mech. Phys. Solids 48, 581–607.

Réti, T., Felde, I., 1999. A non-linear extension of the additivity rule. Comput. Mater. Sci. 15, 466–482.

Sachs, G., 1928. Zur Ableitung einer Fliessdedingung. Z. Ver. Deutsch Ing. 72, 734–736.

Schajer, G., 1988. Measurement of non-uniform residual stresses using the hole-drilling method. Part 1 - stress calculation procedure, and part 2 - practical application of the integral method. Trans. ASME 110, 338–349.

Scheil, E., 1935. Anlaufzeit der Austenitumwandlung. Archiv für Eisenhüttenwesen 12, 565–567.

Schlueter, M., Gerdts, M., Rückmann, J., 2012. A numerical study of MIDACO on 100 MINLP benchmarks. Optimization (Taylor & Francis) 61 (7), 873–900.

Schröder, R., 1985. Untersuchung zur Spannungs- und Eigenspannungsausbildung beim Abschrecken von Stahlzylindern. TH Karlsruhe, Dissertation .

Schwenk, M., 2012. Numerische Modellierung der induktiven Ein- und Zweifrequenzrandschichthärtung. Dissertation, Karlsruhe Institute of Technology KIT Scientific Publishing.

Schwenk, M., Kaufmann, B., Hoffmeister, J., Schulze, V., 2012. Residual stress prediciton for dual frequency induction hardening considering transformation plasticity during austenitization. Proc. 6th Int. Quenching and Control of Distortion Conference , 45–46.

Seok, H.-H., Mun, J.-C., Kang, C.-g., 2015. Micro-crack in zinc coating layer on boron steel sheet in hot deep drawing process. Int. J. Prec. Eng. Manuf. 16, 5, 919–927.

Seth, B. R., 1964. Generalized strain measure with applications to physical problems. Proc. IUTAM Symp. Second-Order Effects in Elasticity, Plasticity and Fluid Dynamics , 162–172.

Shi, D., Hu, P., Liang, Y., 2015. Comparative study of ductile fracture prediction of 22MnB5 steel in hot stamping process. Int. J. Adv. Manuf. Technol. 107, 1–12.

Shi, Z., Dong, J., Ma, W., 2012. Influence of annealing temperature on microstructure and properties of the hot-stamping boron steel. Adv. Mater. Res. 602-604, 385–389.

Sierra, R., Nemes, J., 2008. Investigation of the mechanical behaviour of multi-phase TRIP steels using finite element methods. Int. J. Mech. Sci. 50 (4), 649–665.

Simo, J., Hughes, T., 1998. Computational Inelasticity. Springer, New York.

Sjöström, S., 1985. Interactions and constitutive models for calculation quench stresses on steel. Mater. Sci. Technol. 1, 823–829.

Starink, M., 2001. On the meaning of the impingement parameter in kinetic equation for nucleation and growth reactions. J. Mater. Sci. 36, 4433–4441.

Starink, M., Zahra, A., 1998. Kinetics of isothermal and non-isothermal precipitation in an Al-6at%Si alloy. Philos. Mag. A 77, 187–199.

Steinbeiss, H., So, H., Michelitsch, T., Hoffmann, H., 2007. Method for optimizing the cooling design of hot stamping tools. Prod. Eng. 1 (2), 149–155.

Steinhoff, K., Schupfer, M., Ademaj, A., Weidig, U., 2012. All about press hardening - an overview on technology and markets proceedings. 2nd Int. Seminar Hot Sheet Metal Forming of High-Performance Steel 228, 64–89.

Stopp, R., Schaller, L., Lamprecht, K., Keupp, G., Deinzer, G., 2007. Warmblechumformen in der Automobil-Serienfertigung - Status, Trends, Potenziale. Proc. 2nd Erlanger Workshop Warmblechumformung , 23–36.

Suehiro, M., Sato, K., Tsukano, Y., Yada, H., Senuma, T., Matsumura, Y., 1987. Computer modeling of microstructural change and strength of low carbon steel in hot strip rolling. Trans. Iron Steel Inst. Jpn. 27, 439–445.

Sun, C., Zhang, H., 2013. Microstructural evolution and quenching properties of 22MnB5 steel for hot stamping during resistance heating. Adv. Mater. Res. 849, 75–80.

Suquet, P., 1995. Overall properties of nonlinear composites: a modified secant moduli theory and its link with ponte castañeda's non linear variational procedure. C. R. Acad. Sci. Paris II 320, 563–571.

Swift, M., 1952. Plastic instability under plane stress. J. Mech. Phys. Solids 1, 1–18.

Tabor, D., 1951. The Hardness of Metals. Oxford University Press, London.

Talbot, D., Willis, J., 1992. Some explicit bounds for the overall behaviour of nonlinear composites. Int. J. Solids Struct. 29, 1981–1987.

Taleb, L., Sidoroff, F., 2003. A micromechanical modeling of the Greenwood-Johnson mechanism in transformation induced plasticity. Int. J. Plast. 19, 1821–1842.

Tang, B., Bruschi, S., Ghiotti, A., Bariani, P., 2014. Numerical modelling of the tailored tempering process applied to 22MnB5 sheets. Finite Elem. Anal. Des. 81, 69–81.

Tang, B., Yuan, Z., Cheng, G., Huang, L., Zheng, W., Xie, H., 2013. Experimental verification of tailor welded joining partners for hot stamping and analytical modeling of TWBs rheological constitutive in austenitic state. Mater. Sci. Eng. A 585, 304–318.

Tang, B., Zhao, Z., Yu, S., Chen, J., Ruan, X., 2007. One-step fem based control of weld line movement for tailor-welded blanks forming. J. Mater. Process. Technol. 187-188, 383–386.

Taylor, G., 1938. Plastic strain in metals. J. Inst. Metals 62, 307–324.

Taylor, G., Quinney, H., 1934. The latent energy remaining in a metal after cold working. Proc. Roy. Soc. A 143, 307–326.

Teply, J., Dvorak, G., 1988. Bounds on overall instantaneous properties of elastic-plastic composites. J. Mech. Phys. Solids 36, 29–58.

Terada, K., Kikuchi, N., 2001. A class of general algorithms for multi-scale analysis of heterogeneous media. Comput. Meth. Appl. Mech. Eng. 190, 5427–5464.

thyssenkrupp, February 2016. Thyssenkrupp Steel. `https://www.thyssenkrupp-steel.com/en/startpage.html`.

Todinov, M., 1998. Alternative approach to the problem of additivity. Metall. Mater. Trans. B 29B, 269–273.

Todinov, M., 2000. On some limitations of the Johnson-Mehl-Avrami-Kolmogorov equation. Acta Mater. 48 (17), 4217–4224.

Todinov, M., Knott, J., Strangwood, M., 1996. An assessment of the influence of complex stress states on martensite start temperature. Acta. Mater. 44, 2, 4909–4915.

Torquato, S., 2002. Random Heterogeneous Materials. Microstructure and Macroscopic Properties. Springer-Verlag.

Turetta, A., Bruschi, S., Ghiotti, A., 2006. Investigation of 22MnB5 formability in hot stamping operations. J. Mater. Process. Technol. 177 (1-3), 396–400.

Turteltaub, S., Suiker, A., 2005. Transformation-induced plasticity in ferrous alloys. J. Mech. Phys. Solids 53 (8), 1747–1788.

United States Steel Company Research Laboratory, 1963. Atlas of isothermal transformation diagrams. U.S. Steel Inc., Pittsburgh.

Uppaluri, R., Helm, D., 2016. Thermomechanical characterization of 22MnB5 steels with special emphasis on stress relaxation and creep behavior. Mater. Sci. Eng. A 658 (21), 301–308.

Vermeulen, W., van der Zwaag, S., 1996. Prediction of martensite start temperature using artificial neural networks. Ironmak. and Steelmak. 23 (5), 433–437.

Visintin, A., 1987. Mathematical models of solid-solid phase transitions in steel. IMA J. Appl. Math. 39, 143–157.

Voce, E., 1948. The relationship between stress and strain for homogeneous deformation. J. Int. Metals 74, 537–562.

Voce, E., 1955. A practical strain-hardening function. Metall. 51, 219–226.

voestalpine, February 2016. voestalpine Steel Division. `http://www.voestalpine.com/thinkzinc/phs-ultraform-R`.

Voigt, W., 1889. Über die Beziehung zwischen den beiden Elastizitätskonstanten isotroper Körper. Wied. Ann. 38, 573–587.

Walker KP, Freed AD, J. E., 1994. Thermoviscoplastic analysis of fibrous periodic composites by the use of triangular subvolumes. Compos. Sci. Technol. 50 (1), 71–84.

Wang, J., van der Wolk, P., van der Zwaag, S., 2000. Determination of martensite start temperature in engineering steels part i empirical relations describing the effect of steel chemistry. Mater. Trans. 41, 122–128.

Wang, X., Xu, B., Yue, Z., 2008. Micromechanical modelling of the effect of plastic deformation on the mechanical behaviour in pseudoelastic shape memory alloys. Int. J. Plast. 24, 1307–1332.

Wang, Z., Liu, P., Xu, Y., Wang, Y., Zhang, Y., 2014. Hot stamping of high strength steel with tailored properties by two methods. Procedia Eng. 81, 1725–1730.

Wassermann, G., 1935. Über den Mechanismus der α-γ-Umwandlung des Eisens. Mitt. K.-Wilh.-Inst. Eisenforschung 17, 149–155.

Watt, D., Coon, L., Bibby, M., Goldak, J., Henwood, C., 1988. An algorithm for modelling microstructural development in weld heat-affected zones (part a) reaction kinetics. Acta Metall. 36 (11), 3029–3035.

Willis, J., 1977. Bounds and self-consistent estimates for the overall properties of anisotropic composites. J. Mech. Phys. Solids 25, 185–202.

Willis, J., 1983. The overall response of composite materials. ASME J. Appl. Mech. 50, 1202–1209.

Wolff, M., Böhm, M., Dalgic, M., Hüßler, I., 2008a. Evaluation of models for TRIP and stress-dependent transformation behaviour for the martensitic transformation of the steel 100Cr6. Comput. Mater. Sci. 43, 108–114.

Wolff, M., Böhm, M., Helm, D., 2008b. Material behavior of steel- modeling of complex phenomena and thermodynamic consistency. Int. J. Plast. 24, 746–774.

Wriggers, P., 2008. Nonlinear Finite Element Methods. Springer.

Wu, W., Hu, P., Shen, G., 2015. Thermomechanical-phase transformation simulation of high-strength steel in hot stamping. Math. Probl. Eng. 2015, 12 p.

Xiao, Y., Chen, J., Cao, J., 2012. A generalized thermodynamic approach for modeling nonlinear hardening behaviors. Int. J. Plast. 38, 102–122.

Yanagimoto, J., Oyamada, K., 2005. Springback of high-strength steels after hot and warm sheet formings. CIRP Ann. - Manuf. Technol. 54, 1, 213–216.

Yang, H., Bhadeshia, H., 2009. Austenite grain size and the martensite-start temperature. Scripta Mater. 60 (11), 493–495.

Yoshie, A., Fujita, T., Fujioka, M., Okamoto, K., Morikawa, H., 1996. Formulation of flow stress of Nb steels by considering work-hardening and dynamic recovery. ISIJ Int. 36 (4), 467–473.

Yu, C., Kang, G., Kan, Q., 2014a. Crystal plasticity based constitutive model of NiTi shape memory alloy considering different mechanisms of inelastic deformation. Int. J. Plast. 54, 132–162.

Yu, C., Kang, G., Kan, Q., 2014b. Crystal plasticity based constitutive model of NiTi shape memory alloy considering different mechanisms of inelastic deformation. Int. J. Plast. 54, 132–162.

Yu, H., 1977. Berechnung von Abkühlungs-, Umwandlungs-, Schweiss-, sowie Verformungseigenspannungen mit Hilfe der Methode der finiten Elemente. TH Karlsruhe, Fak. Maschinenbau, Dissertation .

Zeller, R., Dederichs, P. H., 1973. Elastic constants of polycrystals. Phys. Stat. Sol.(B) 55, 831–842.

Zener, C., 1946. Kinetics of the decomposition of austenite. Trans. Am. Inst. Min. Metall. Eng. 167, 550–595.

Zieliński, A., Frey, F., 2001. On linearization in non-linear structural finite element analysis. Comput. Struct. 79, 825–838.

Zwick Roell, February 2016. Universal testing machines from 1 kN to 250 kN. `http://www.zwick.co.uk/en/products/static-materials-testing-machines.html`.

Schriftenreihe Kontinuumsmechanik im Maschinenbau Karlsruher Institut für Technologie (KIT) (ISSN 2192-693X)

Herausgeber: Prof. Dr.-Ing. Thomas Böhlke

Band 1 Felix Fritzen
Microstructural modeling and computational homogenization of the physically linear and nonlinear constitutive behavior of micro-heterogeneous materials. 2011
ISBN 978-3-86644-699-1

Band 2 Rumena Tsotsova
Texturbasierte Modellierung anisotroper Fließpotentiale. 2012
ISBN 978-3-86644-764-6

Band 3 Johannes Wippler
Micromechanical finite element simulations of crack propagation in silicon nitride. 2012
ISBN 978-3-86644-818-6

Band 4 Katja Jöchen
Homogenization of the linear and non-linear mechanical behavior of polycrystals. 2013
ISBN 978-3-86644-971-8

Band 5 Stephan Wulfinghoff
Numerically Efficient Gradient Crystal Plasticity with a Grain Boundary Yield Criterion and Dislocation-based Work-Hardening. 2014
ISBN 978-3-7315-0245-6

Band 6 Viktor Müller
Micromechanical modeling of short-fiber reinforced composites. 2016
ISBN 978-3-7315-0454-2

Die Bände sind unter www.ksp.kit.edu als PDF frei verfügbar oder als Druckausgabe bestellbar.

Band 7 Florian Rieger
Work-hardening of dual-phase steel. 2016
ISBN 978-3-7315-0513-6

Band 8 Vedran Glavas
Micromechanical Modeling and Simulation of Forming Processes. 2017
ISBN 978-3-7315-0602-7

Band 9 Eric Bayerschen
Single-crystal gradient plasticity with an accumulated plastic slip: Theory and applications. 2017
ISBN 978-3-7315-0606-5

Band 10 Bartholomäus Brylka
Charakterisierung und Modellierung der Steifigkeit von langfaserverstärktem Polypropylen. 2017
ISBN 978-3-7315-0680-5

Band 11 Rudolf Neumann
Two-Scale Thermomechanical Simulation of Hot Stamping. 2017
ISBN 978-3-7315-0714-7